Exergy - New Technologies and Applications

Edited by Kenneth Eloghene Okedu

Published in London, United Kingdom

Exergy - New Technologies and Applications
http://dx.doi.org/10.5772/intechopen.104062
Edited by Kenneth Eloghene Okedu

Contributors
Ismaila Badmus, José Luis Velázquez Ortega, Aldo Gómez López, Masoumeh Bararzadeh Ledari, Reza Bararzadeh Ledari, Mohammad Saleh Barghi Jahromi, Abdallah A. Ibrahim, Loay E. George, Ali Raza, Muhammad Khurram, Liaquat Ali Khan, Zunaira-Tu Zehra, Muhammad Ahsan Pervaiz Khan, Asif Durez, Sana Toumi, Mohamed Benbouzid, Mohamed Faouzi Mimouni, Andrei Sergeev, Kimberly Sablon

First published in London, United Kingdom, 2023 by IntechOpen
IntechOpen is the global imprint of INTECHOPEN LIMITED, registered in England and Wales, registration number: 11086078, 5 Princes Gate Court, London, SW7 2QJ, United Kingdom

British Library Cataloguing-in-Publication Data
A catalogue record for this book is available from the British Library

Additional hard and PDF copies can be obtained from orders@intechopen.com

Exergy - New Technologies and Applications
Edited by Kenneth Eloghene Okedu
p. cm.
Print ISBN 978-1-83768-544-8
Online ISBN 978-1-83768-545-5
eBook (PDF) ISBN 978-1-83768-546-2

Meet the editor

Kenneth Eloghene Okedu was a research fellow in the Department of Electrical and Computer Engineering, Massachusetts Institute of Technology (MIT), Boston, USA, in 2013. He obtained his Ph.D. from the Department of Electrical and Electronic Engineering, Kitami Institute of Technology, Japan, in 2012. He received his BSc and MEng in Electrical and Electronic Engineering from the University of Port Harcourt, Nigeria, in 2003 and 2007, respectively, where he has been a faculty member since 2005. He has also been a visiting faculty member at the Abu Dhabi National Oil Company (ADNOC) Petroleum Institute and at the Caledonian College of Engineering, Glasgow Caledonian University, UK. He is presently a visiting professor in the Department of Electrical and Computer Engineering, National University of Science and Technology (NUST), Oman, and an adjunct professor in the Department of Electrical and Electronic Engineering, Nisantasi University, Turkey. He was recognized among the top 1% of peer reviewers in engineering by Publons (Web of Science) in 2018 and 2019 and was the editor's pick in the *Journal of Renewable and Sustainable Energy* in 2018. Dr. Okedu has published several books and journals/transactions on renewable energy. He is an editor for several journals, including *Frontiers in Renewable Energy Research* (Smart Grids), *Energies*, *International Journal of Smart Grids*, *International Journal of Renewable Energy Research*, *International Journal of Electrical Engineering*, *Mathematical Problems in Engineering*, and *Trends in Renewable Energy*. His research interests include power system stability, renewable energy systems, stabilization of wind farms, stability analysis of doubly fed induction generators (DFIGs) and permanent magnet synchronous generators (PMSG), variable-speed wind turbines, augmentation and integration of renewable energy into power systems, grid frequency dynamics, wind energy penetration, FACTS devices and power electronics, renewable energy storage systems, and hydrogen and fuel cells. Dr. Okedu was listed among the top 2% of scientists by Stanford University as well as the top 2% of the most academically cited researchers worldwide in the Scopus Worldwide Database. He received the Outstanding Publication Award for publishing the most Scopus-indexed papers in 2021–2022 at the National University of Science and Technology (NUST), Oman. He is currently a researcher with the Smart Energy Unit, Victoria University, Melbourne, Australia, and a teaching faculty of graduate and post-graduate programs at the Melbourne Institute of Technology, Australia.

Contents

Preface

The concept of exergy is very important in engineering and science analyses. The useful amount of work extracted from any system could be obtained using the principles of exergy considering the equilibrium of the system with the surrounding environment in a reversible state. This book is divided into two sections, each of which includes four chapters: "New Technologies in Exergy" and "Applications of Exergy".

In Chapter 1, drying is emphasized as an energy-intensive process in food preservation because it eliminates extra moisture and increases the longevity of food products. The chapter examines the use of both renewable and non-renewable energy sources to generate the energy needed for drying. It highlights the merits of using solar energy as the best source of energy for drying processes, including the drying of agricultural produce. The main components of a typical solar dryer are the fan, the solar air heater (SAH), and the dryer chamber. The chapter also carries out an exergy economic analysis to ensure that the primary contributing factors to system exergy loss are recognized and understood.

Compressing images and reconstructing them without degrading their original quality is a major challenge in our modern world. Thus, Chapter 2 proposes a coding system that considers the implementation of both quality and compression rate considering high synthetic entropy coding schema. This coding schema can store the compressed image at the smallest size possible without affecting its original quality. The chapter considers a technique based on Discrete Cosine Transform (DCT) and Discrete Wavelet Transform (DWT). The evaluation of the test case was carried out using different standard color images and various performance metrics of the variables to reflect the effectiveness of the proposed scheme.

The approach of nonreciprocal photonic management can turn the absorption-emission balance to the advantage of absorption and improve the conversion efficiency more than the usual Shockley–Queisser balanced limit. Chapter 3 examines the use of nonreciprocal photovoltaic (PV) cells to convert the entire exergy (Helmholtz free energy) of quasi-monochromatic radiation into electric power. It presents the evaluation of the limiting performance of a typical nonreciprocal, dissipation-less monochromatic converter and discusses the limiting efficiency of the nonreciprocal converter based on the adverse effect of the greenhouse. The chapter also presents a thorough modeling of the greenhouse effect in the gallium arsenide (GaAs) PV converter along with the performance of a PV in converting laser radiation. The presented results show that the greenhouse filter ensured a sharp absorption edge and reduced conversion losses related to the distributed PV bandgap and laser-cell matching losses.

In a bid to replace fossil fuels with renewable energy resources, biofuels have emerged as potential replacements for diesel fuels. These biofuels can be used in an engine cylinder without any modifications by blending them with conventional diesel fuels. Evaporation of fuel droplets in engine cylinders is a process involving the atomization of fuel to release gases. This evaporation process is directly connected to the efficiency of the engine. Chapter 4 presents the evaporation characteristics of conventional diesel fuel and biofuels under different working conditions. It also models the evaporation phenomenon using computational fluid dynamics (CFD) and the effects of cylinder conditions. The results show that biofuel droplets have a better evaporation rate at high operating conditions in an engine cylinder.

Chapter 5 presents the dynamic behavior of a non-commercial shrink tank heat transfer process, considering two different arrangements of electrical resistances for water heating. Thorough numerical simulations are carried out based on the implicit method of alternating directions (ADI). The evaluation of the system's performance based on heating times with their respective temperature distributions shows that the arrangement with four resistances is the most efficient for the process of heating the water in the shrink tank.

Chapter 6 estimates entropy generation in terrestrial systems and the atmosphere by imitating the entropy analysis of steam power generation (STPG). The maximum entropy generation is related to the outgoing longwave radiation flux. The presented results show that the most significant terms of entropy generation (heat dissipation) in different processes are related to the hidden and sensible heat fluxes. It is observed that the vegetation cover (boiler system) destroyed a part of solar energy absorption in the form of entropy generated by the formation of water vapor and transpiration (steam turbine).

Chapter 7 presents the state of the art of marine current turbine (MCT) system faults. The MCT topology consists of a marine turbine, a generator (permanent magnet synchronous generator (PMSG) or doubly fed induction generator (DFIG)) and a PWM power converter. This chapter explores the several faults related to the turbine, generator, blades, and converters in an MCT. These faults generate oscillations in the speed and the torque, which may lead to mechanical vibrations and the rapid destruction of the insulating material generator.

Chapter 8 examines the importance of exergy-based parameters like exergy efficiency, environmental compatibility, sustainability index, and depletion number, as well as the Improvement Potential (IP) of hydrocarbon fuel utilization. The IP of hydrocarbon fuel utilization is the product of the square of the depletion number and the fuel input exergy. Various metrics are used to evaluate the fuel exergy performance in the presented study. Due to the fact that exergy destruction is linked to material depletion, the fuel exergy flow is therefore proportional to its material flow. This chapter shows that if the IP, is high, it means that the exergy losses are too high and there is a big room for exergy efficiency improvement.

Kenneth Eloghene Okedu
Smart Energy Unit,
Victoria University,
Melbourne, Australia

Melbourne Institute of Technology,
School of Information Technology and Engineering,
Melbourne, Australia

Section 1

New Technologies in Exergy

Chapter 1

Exergy of Solar Dryer

Mohammad Saleh Barghi Jahromi

Abstract

Due to the fact that it eliminates extra moisture and increases food products' shelf lives, drying is an energy-intensive process in food preservation. Both renewable and non-renewable energy sources can be used to generate the energy needed for drying. Researchers have recently given sources like solar energy the highest consideration when employing renewable energy. Solar energy is the best source of energy for the drying process with solar dryer systems because it is free, clean, available, and economically viable. The usage of solar dryers in agricultural production areas like farms and gardens conserves a variety of energy resources (such as fossil fuel), improves food-processing efficiency, and lowers the cost of transportation. The main components of solar dryers are the fan, the solar air heater (SAH), and the dryer chamber, which is why there are different exergy factors. In the industry of solar dryers, it is crucial to improve drying energy effectiveness and lower energy consumption costs. Using modern technologies makes it easier to improve energy efficiency and lower operational expenses. The main goal of many studies today is to evaluate the energy costs of various drying techniques. This technique, also known as exergy economic analysis, makes sure that the primary contributing factors to system exergy loss are recognized and understood.

Keywords: solar dryer, exergy efficiency, exergy loss, renewable energy, drying techniques

1. Introduction

Most countries in the world are facing the problem of running out of fossil fuels; many scientists are using renewable energy to solve this problem. Therefore, it is very important to use systems that work with renewable energy. One of these methods is the use of solar dryers, which have different types and are capable of drying all kinds of products and medicinal plants, and help the country's economy a lot. To increase product quality, reduce product waste in gardens and packaging industries, and use dried products in seasons where fresh products are not available, drying with a solar dryer is the best method. A significant portion of the moisture content of the items is removed during the dehumidification process, which also significantly lowers the activity of microorganisms during the storage time. Open sun drying (OSD) is an easy method with a very low cost. With this method, most agricultural products can be dried and it is an effective method in most countries. However, it is impossible to manage the factors that effect drying, including humidity, temperature, mass flow rate for drying, and heat entering the drier chamber. As a result, it causes an unfavorable drying speed or a longer drying time. The OSD approach has additional drawbacks, such as lowering product quality due to wind, trash, rain, insects, and animals [1, 2]. Fossil fuel combustion often produces hot air for industrial drying, and huge amounts of energy (about 13% in the agricultural industry) are used globally for this purpose [3]. The quality of the dried product is one of the key factors in the drying process. Therefore, the hot air used to dry the product should be in the range of 45 to 60°C. The intensity of solar radiation decreases during sunset hours and the process of drying and dehumidifying the product continues in the dryer chamber, so dryers must dry the product continuously. As it was said, the quality of the dried product is important, so the temperature inside the drying chamber should not be too high, also at noon, when the solar radiation increases. The optimum option might be the thermal storage system (TES), according to the literature review. PCMs (phase change materials), can be used in the solar dryer system, the most important PCM can be mentioned paraffin wax. Numerous studies based on solar dryers of various technologies have been carried out, and the majority of them compared the results of a solar-assisted drying method with a conventional method. When compared to the conventional drying method, drying wet materials with solar dryers reduces drying time while improving dried sample quality [1–3]. **Table 1** shows new studies on the technology of PCM-equipped solar dryers. In addition, the key results of each research are given separately. The use of heat during the drying process could result in significant changes in the product quality. Low temperatures are thus preferred, particularly for drying medical plants, fruits, and vegetables. Low temperature necessitates the use of a high-quality energy source,

Reference	Collector type and PCM storage location	Dried product	Important results
[4]	Flat plate solar collector (FPSC) - Paraffin RT-42 is located inside the dryer chamber.	Valeriana jatamansi medicinal plant	Using this dryer, Valeriana jatamansi medicinal plant was dried to good quality and the final moisture content of the product reached 9% in 5 days. The use of PCM reduced the drying time and made the temperature of the dryer chamber uniform in the evening hours. By connecting this dryer to the heat pump, the drying time lasted 8 days and with the OSD method, it took 14 days.

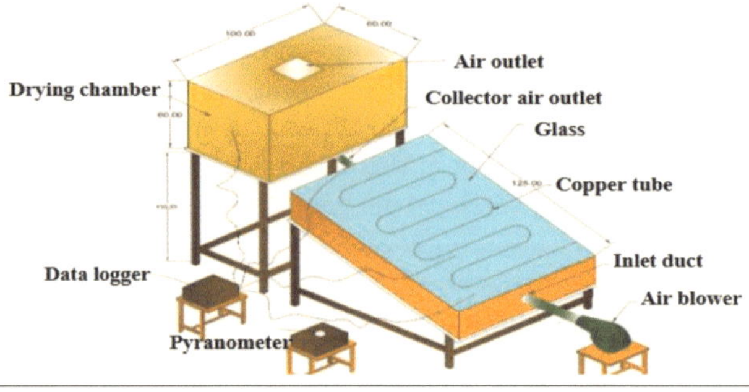

Reference	Collector type and PCM storage location	Dried product	Important results
[5]	The evacuated tube solar collector (ETSC) - paraffin RT50 is located in the copper coil, inside the expansion source.	Apple	With a mass flow rate of 0.05 kg/s, the storage system used with PCM improves internal thermal energy by 12.5 MJ. The internal heat energy is decreased by 1.72 MJ by decreasing the air mass flow rate by 0.025 kg/s.
	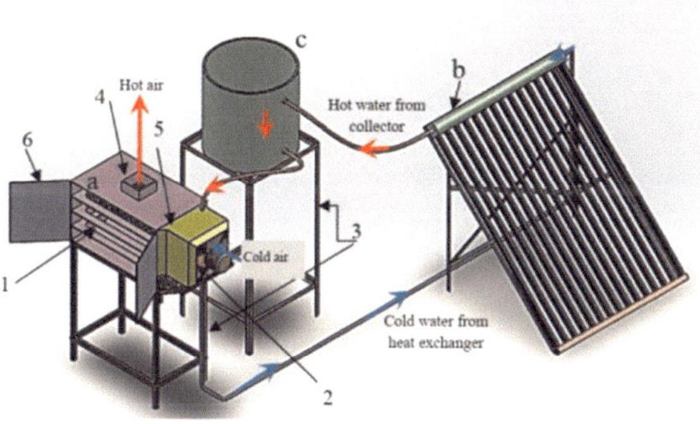		
[6]	Parabolic Trough Solar Collector (PTSC) - paraffin wax is located in the copper coil in the expansion tank.	Apple	The fluids used in this test include PCM, engine oil, glycerin, and water, which are located in the copper pipes located in the expansion source. The highest temperature is 83.8, 88.2, 85.6, and 79.2 degrees Celsius for Nano fluid, oil, glycerin, and water, respectively. The efficiency of the collector is in the range of 71.9 to 73.4%.
	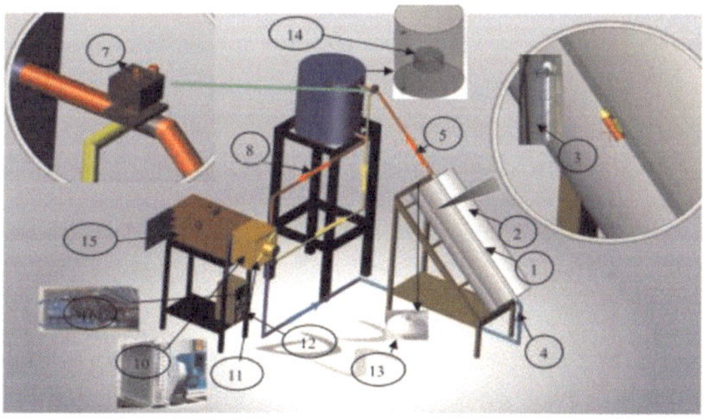		
[7]	Solar air heater (SAH) - paraffin wax is located under the absorber plate.	banana	Banana slices were to be dried for 18 hours, of which 13 hours were spent during the day and 5 hours were spent draining PCM material. - Average drying efficiency was found to be 2.98%, and collector efficiency to be 66.32%.

Reference	Collector type and PCM storage location	Dried product	Important results
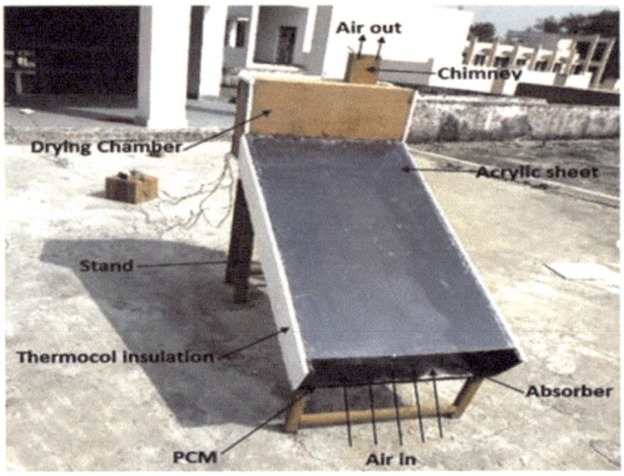			

Table 1.
A review of new technologies of solar dryers equipped with PCM.

which significantly raises irreversibility [2]. Therefore, it is essential to increase drying energy efficiency and reduce the costs related to energy use in this sector. Increasing energy efficiency and reducing operating costs is more possible with the use of new technologies. For this reason, recent studies have focused on increasing energy efficiency and reducing the operating costs of drying systems with renewable energy sources. Energy cost evaluations for various drying methods are the primary focus of many studies today. Exergy economic analysis, another name for this method, ensures that the main causes of the system's exergy loss are clearly identified. This makes it possible to pinpoint the system component that requires improvement and to put the right methods into action to boost the exergy efficiency.

2. Solar dryers

Solar dryers are divided into two categories: forced convection (FC) and natural convection (NC) based on the air circulation employed for drying [8, 9]. As a result, altering the drying method alters the characteristics of agricultural samples during the drying process. The hot air required in the drying process is supplied to the drying chamber through a solar air heater using an external device such as a fan. Forced circulation solar dryers are also known as active solar dryers. The airflow needed to dry the samples in natural circulation dryers (or inactive solar dryers) is caused by

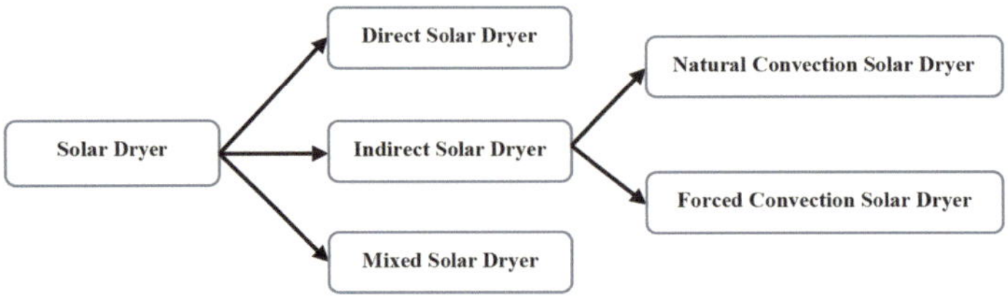

Figure 1.
Classification of solar dryers and drying techniques [10].

gravity or buoyant force. **Figure 1** depicts the classification of solar dryers with natural and forced circulation into three categories: direct, indirect, and mixed-mode. Based on the ways of drying displayed in **Figure 1**, a list of the solar dryers available for drying agricultural commodities [10–12].

The direct, indirect, and mixed-mode categories of solar dryers are described in the preceding section [5, 10]. Dryer chambers are present in direct sun dryers. This chamber is comprised of an opaque cover and is insulated. Usually, the drying chamber is made of materials such as glass, plastic, and galvanized sheets, and in order for air to enter the drying chamber and the product to dry, holes are placed at the entrance and exit of the chamber [13, 14]. Natural sun-drying (A1), Natural rack or shade drying (A2), Staircase solar drying (A3), and Foldable solar drying (A4) are examples of direct solar drying (**Figure 2**), which can also be seen that the product is exposed to direct sunlight, so the quality of the dried products is low and

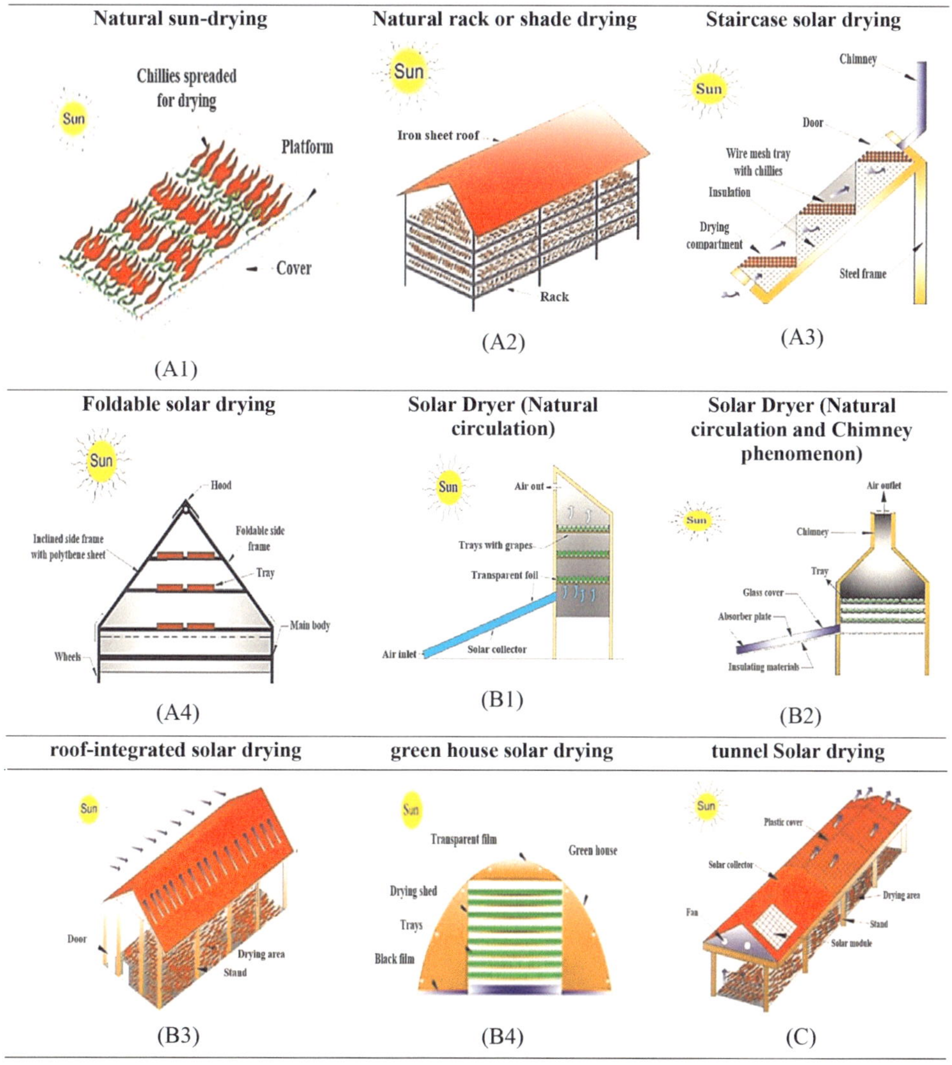

Figure 2.
Classification of types of solar dryers for drying [10].

Environmental parameters cannot be controlled [10, 15]. One of the advantages of direct solar dryers is that they are economical because they are easy to make and have a low cost. This dryer can protect the product from dust, wind, and rain. The main disadvantages of these dryers can be mentioned as the lack of control of environmental factors such as radiation intensity, speed, temperature entering the dryer chamber, and long drying time and reducing the quality of the dried product [16]. To overcome the disadvantages of direct solar dryers, the hot air needed to dry the product was provided by solar air heaters. Therefore, in indirect solar dryers, sunlight does not shine directly on the product inside the dryer chamber. A blower or suction fan can be added to indirect solar dryers to create a forced convection to dry the product [17]. Higher temperature values can be attained with regulated airflow rates using the solar collector unit [18]. The speed of air movement in the solar collector, which can be managed by a fan or blower, affects the effectiveness of an indirect solar drier. A fan's principal function in a dryer unit is to maintain the correct airflow rate, which promotes even moisture removal from products [19]. Solar Dryers with Natural circulation (B1), with Natural circulation and Chimney phenomenon (B2), Integrated roof (B3), and greenhouse (B4) are examples of indirect solar dryers (**Figure 2**), whose common feature is the supply of hot air by a solar air heater, which at the end it is connected to the dryer chamber [10]. In mixed-mode solar dryers, due to the simultaneous use of the solar air heater and the drying chamber against direct sunlight, the drying efficiency raises and reduces the drying time of the product. Solar drying tunnel (C) and hybrid dryers belong to the mixed-mode category [10].

3. Calculation of exergy in solar dryers

There are various components in solar dryers, such as solar collector, fan, and dryer chamber, so many factors may play a role in exergy due to the presence of these components in the dryer system. The optimization of each of these components is determined using exergy analysis and one of these parameters is the calculation of exergy loss. According to the results obtained from exergy analysis, suction fans in solar dryers have the highest exergy destruction and the lowest exergy efficiency [2, 20].

Important factors in the drying process can be mentioned sunlight, the mass flow rate of air, and humidity inside the drying chamber. Exergy loss decreases with the increase of air mass flow rate, so the effect of humidity on exergy changes is insignificant [21]. Solar dryers have rather poor exergy efficiency, according to an examination of energy and exergy are done on them [22]. The calculations do not consider the effects of kinetic and potential energy, and the exergy balance of a product can be calculated from Eq. (1):

$$E_i - E_o = E_{dest} \tag{1}$$

The $\boldsymbol{E_{dest}}$, which indicates the destructive exergy for the drier throughout the drying phase of the slices, can be represented as (Eq. 2):

$$E_{dest} = Ex_{in} + \mathrm{W} - Ex_{u,p} \tag{2}$$

$Ex_{u,p}$ is the real exergy in Eq. (2) and may be determined using the pressure drop in the collector:

$$Ex_{u,p} = Ex_u - Ex_{\mathrm{w}} \tag{3}$$

Input (Ex_{in}) and output exergy (Ex_u) parameters can be calculated from Eqs. (4) and (5) [2]:

$$Ex_{in} = \dot{m}c_p\left[(T_o - T_i) - T_a\left(Ln\frac{T_o}{T_i}\right)\right] \quad (4)$$

$$Ex_u = \dot{m}\left[c_p(T_o - T_i) - T_a\left(c_v Ln\left(\frac{T_o}{T_i}\right) - RLn\left(\frac{\rho_{out}}{\rho_{in}}\right)\right)\right] \quad (5)$$

The dryer's pump and fan are tied to Ex_W, which represents the system's additional source of energy.

Ex_W For the pump was estimated using a wattmeter, while Ex_W for the fan was computed using equation [2]:

$$Ex_W = \frac{T_a}{T_i}W_{fan}, W_{fan} = \frac{\dot{m} \times \Delta P}{\left(\rho \times \eta_{fan}\right)} \quad (6)$$

η Fan indicated the fan efficiency value, which for the current system was 0.91.

Eq. (7) can be used to determine the exergy efficiency of collectors while taking the sun's temperature (4350 K) and the inlet and output fluid temperatures into account [2, 23]:

$$\eta' = \frac{\dot{m}c_p\left[(T_o - T_i) - T_a\left(ln\,\frac{T_a}{T_i}\right)\right]}{A_c I_o\left[1 - \frac{T_a}{T_{sun}}\right]} \quad (7)$$

Additionally, the results of Eqs. (8) and (9) [24], which were used to calculate the exergy efficiency for the drying chamber and the drying process, are as follows:

$$\eta'_{dry,cab} = \frac{\left[1 - \frac{T_o}{T_{in}}\right]\dot{Q}_{o,dryer}}{Ex_{in}} \quad (8)$$

$$\eta'_{dry} = \frac{\dot{m}c_p\left[(T_o - T_a) - T_a\left(Ln\frac{T_o}{T_a}\right)\right]}{\dot{m}c_p\left[(T_i - T_a) - T_a\left(Ln\frac{T_i}{T_0}\right)\right]} \quad (9)$$

4. A review of the results of articles on exergy solar dryers

In a study, a solar dryer system fitted with PCM underwent an experimental examination and an exergy analysis. **Figure 3a** shows exergy efficiency and exergy loss on the test days. Due to the use of thermal energy storage used in this paraffin wax test, exergy loss occurs less during daylight hours. The results show that the lowest exergy loss with 0.18117 kW and the highest exergy efficiency with 69.59% is related to the first day of the experiment. On the second and third days, exergy loss increases due to the amount of drying of the product and the change in environmental conditions. Economic exergy analysis is shown in **Figure 3b**. It can be concluded that the highest cost of exergy destruction is related to the fans in solar dryers and they have a minimum exergy efficiency of 55.28%. Therefore, the fans in the dryers should be optimized [20].

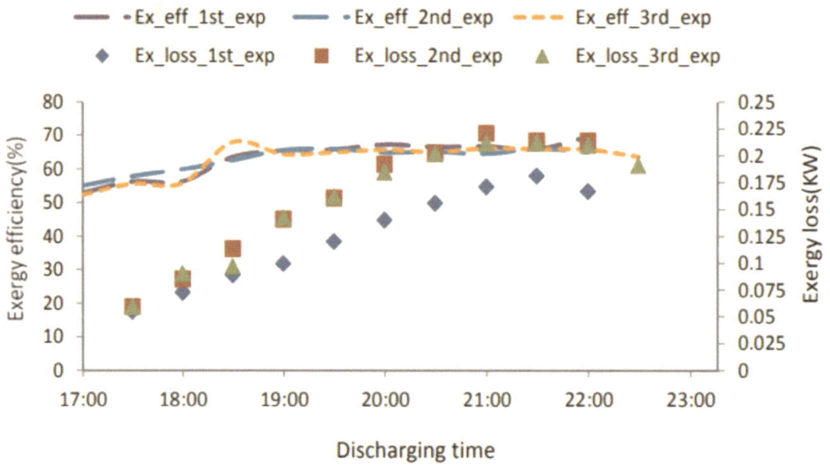

(a)

Component	Ex_{eff} (%)	Ex_f (%)	R_{ex} (MW/$)	Z_{total} ($/h)	C_D ($/h)	C_{total} ($/h)	f_{ex} (-)
Air collectors	75.54	16.34	0,073	0,702	0,0044	0,7064	0,991
Fans	55.96	13.18	0,266	0,235	0.2286	0,4636	0,507
Drying cabin	94.46	17.52	0,042	1.607	0.1080	1,7150	0,937
Heat exchanger	63.49	14.74	0.068	0.430	0.1170	0.5470	0.786
Energy storage medium	60.56	13.89	0.184	0.343	0.1784	0.5214	0.658
Fittings	82.43	15.69	0,102	0,420	0.0083	0,4283	0,980

(b)

Figure 3.
(a) Values of exergy efficiency and exergy loss during the test days, (b) Exergeoeconomic analysis for the solar dryer system in this research [20].

In other study, the exergy efficiency for the drying process Jerusalem Artichoke slices with the dryer integrated with PCM varies between 35.3–59.7% (**Figure 4b**) and, for the system without PCM ranged between 17.1 to 42.9% (**Figure 4b**). This means that PCM usage improved the exergy efficiency (at least 28.62%). In both cases with and without PCM, exergy loss and exergy destruction increase with increasing air mass flow rate (**Figure 4a**) [2]. The specific energy consumption (SEC) of 2.62 kWh/kg was discovered in another study. Additionally, the average energy efficiency of solar drying was 30%, with a range of 1 to 93% (**Figure 5**). Improvement potential values were shown to be between 0.3 and 630 W, with an average of 247 W (**Figure 5**) [25].

For both configurations, the exergy inflow, outflow, and losses were calculated for the collector and drying chamber. In forced and natural convection modes, the exergy outflow of the SAC was between 1.04 and 46.85 W and 1.13 and 50.94 W, respectively (**Figure 6a**). Under forced and natural convection, the exergy loss for the drying chamber ranged from 0.062 to 21.99 W and 0.394 to 24.99 W, respectively (**Figure 6b**). Therefore Average exergy efficiencies for SAC and drying chamber in ISD with forced convection were 2.03 and 59.32%, while these values

	Drying condition		Exergy (kJ/s)	
	Airflow rate (kg/s)	EL	ETR	DE
WPCM	0.025	0.131	0.012	0.009
	0.05	0.188	0.023	0.020
	0.9	0.206	0.032	0.32
WOPCM	0.025	0.142	0.024	0.018
	0.05	0.174	0.030	0.021
	0.9	0.196	0.045	0.030

(a)

	Drying condition	Exergy efficiency (%)	
	Airflow rate (kg/s)	Drying	Cabinet
WPCM	0.025	35.3	46.1
	0.05	47.2	64.4
	0.9	59.7	80.8
WOPCM	0.025	17.1	25.6
	0.05	26.6	31.9
	0.9	42.9	43.6

(b)

Figure 4.
(a) Exergy loss (EL), Exergy input (ETR), and Exergy destruction (DE) parameters for solar dryers equipped with PCM for different mass flow rates, (b) Exergy efficiency values for the cases with and without PCM for the drying chamber and the drying process [2].

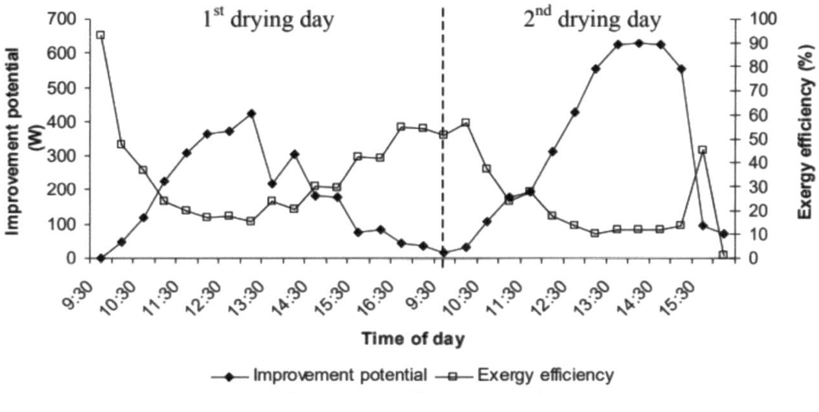

Figure 5.
Changes in recovery potential and exergy efficiency during the test period [25].

were 2.44 and 55.45% in ISD with natural convection [26]. In another study, energy and exergy evaluations were used to determine how well each component of the dewatering system performed. The calculated mean values of the SAC's energy and exergy efficiencies were 9.8 (26.10%) and 0.14 (0.81%), respectively [27].

In another study, an experimental study was conducted on an indirect solar dryer (ISD). The drying process has been carried out at different flow rates. The findings

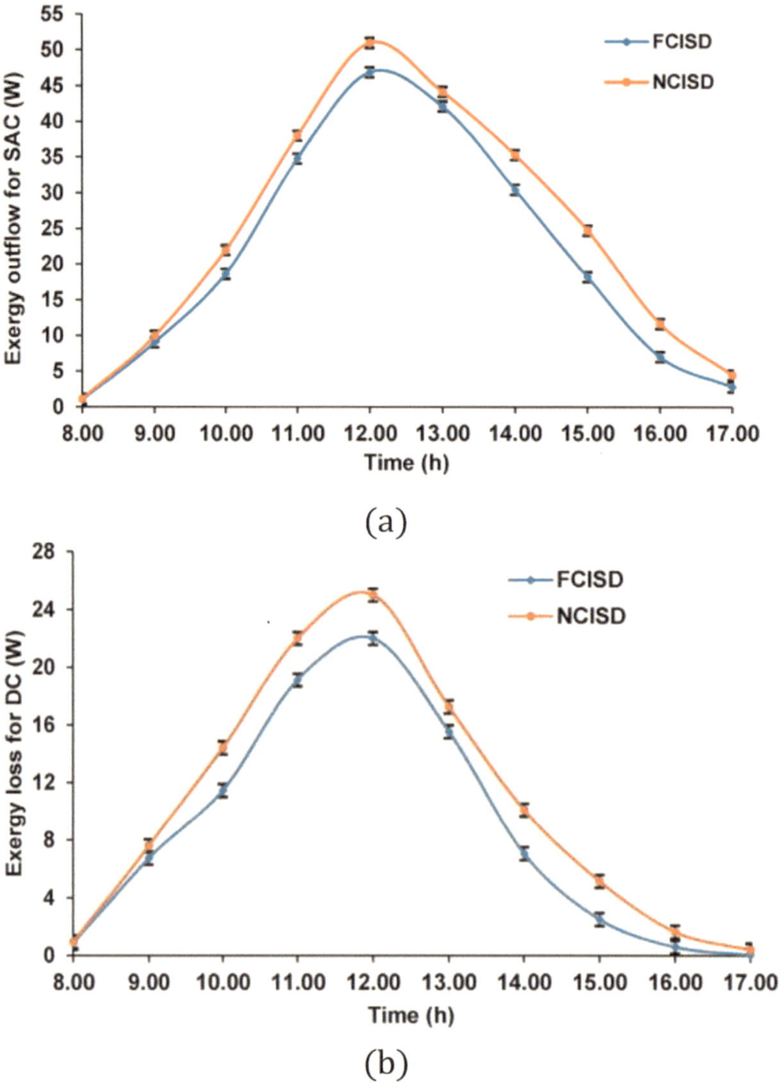

Figure 6.
(a) Exergy output for forced and natural convection from solar air heater (SAH), (b) exergy losses from solar air heater (SAH) [26].

indicate that when air mass flow rate increases, exergy efficiency rises, and exergy losses fall. As shown in **Figure 7**, the energy efficiency rose from 28.74 to 40.67% with the increase in mass flow rate from 0.0141 to 0.0872 kg/s [28].

5. Conclusion

In this article, a review of the mechanism of solar dryers with drying technologies with PCM, exergy calculation formulas, exergy losses, and exergy destruction was

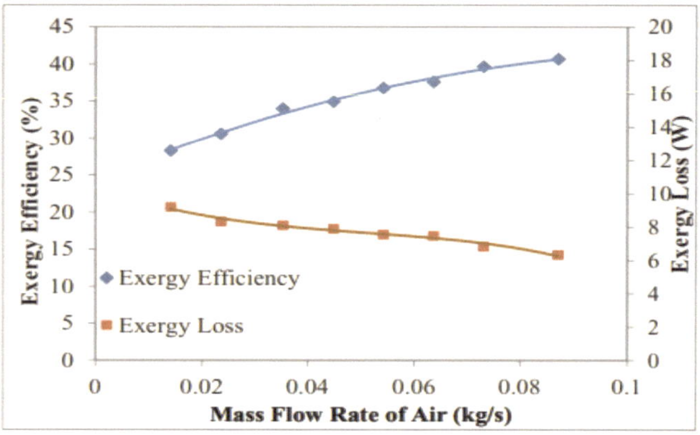

Figure 7.
Exergy losses and exergy efficiency in different mass flow rates during the test period.

investigated. In addition, new articles about the results of exergy analysis and parameters such as exergy efficiency, losses, and exergy destruction for new solar dryers were reviewed, the important results of which are:

- Indirect solar dryers with forced convection have more exergy losses due to the presence of a suction fan and have higher exergy efficiency compared to natural convection.
- The use of PCM in solar dryers significantly improves exergy efficiency. Because the most important problem of solar dryers is the continuous drying of the product in the afternoon until the night, to solve this problem, a thermal storage system is used. The most commonly used PCM is paraffin wax, which is connected to copper pipes inside the storage tank. They are placed in the solar collector or inside copper or aluminum boxes and placed inside the drying chamber or inside the solar air heater.
- With an increase in air mass flow rate, exergy efficiency rises and exergy loss falls [2, 20].
- According to the findings of prior studies, the fans used in solar dryers have the largest exergy destruction and the lowest exergy efficiency.
- Exergy efficiency can be increased by optimizing the suction fans or the air recycling system can be used in solar dryers [2, 29].
- Humidity is one of the important factors in the drying process. The drying process is carried out slowly because the final moisture content of the product must be reached and the product dried, so the volume of the incoming air is small compared to changes in humidity, so the humidity changes in exergy are insignificant.

Nomenclature

$Ex_{u,p}$	exergy outlet considering pressure drop (W)
Ex_{dest}	exergy destruction (W)
Ex_{in}	exergy inlet (W)
Ex_{u}	exergy outlet without considering pressure drop (W)
Ex_{w}	exergy for work (W)
$\dot{m}$	Mass flow rate (kg/s)
T_a	Air temperature (°C)
T_o	Temperature outlet (°C)
T_i	Temperature inlet (°C)
A_c	Collector area (m^2)
I_o	Solar radiation normal to the collector (W/m^2)
$\dot{Q}_{o,dryer}$	Energy output from the chamber dryer (J)
ρ	Density (kg/m^3)

Abbreviations

TES	Thermal energy storage
PCM	Phase change material
SEC	Specific energy consumption
ISD	Indirect solar dryer
SAC	Solar air collector

Author details

Mohammad Saleh Barghi Jahromi
Faculty of Mechanical Engineering, Department of Mechanical Engineering, Yazd University, Iran

*Address all correspondence to: msc.barghi@gmail.com

References

[1] Mokhtarian M, Tavakolipour H, Ashtari AK. Effects of solar drying along with air recycling system on physicochemical and sensory properties of dehydrated pistachio nuts. LWT. 2017;**75**:202-209

[2] Jahromi MSB, Iranmanesh M, Akhijahani HS. Thermo-economic analysis of solar drying of Jerusalem artichoke (Helianthus tuberosus L.) integrated with evacuated tube solar collector and phase change material. Journal of Energy Storage. 2022;**52**: 104688

[3] Barghi Jahromi MS, Iranmanesh M. Experimental investigation on the use of PCM in a pistachio solar dryer by the evacuated heat pipe solar collector. Journal of Pistachio Science and Technology. 2019;**3**(6):73-87

[4] Bhardwaj AK, Chauhan R, Kumar R, Sethi M, Rana A. Experimental investigation of an indirect solar dryer integrated with phase change material for drying valeriana jatamansi (medicinal herb). Case Studies in Thermal Engineering. 2017;**10**:302-314

[5] Iranmanesh M, Akhijahani HS, Jahromi MSB. CFD modeling and evaluation the performance of a solar cabinet dryer equipped with evacuated tube solar collector and thermal storage system. Renewable Energy. 2020;**145**: 1192-1213

[6] El Khadraoui A, Bouadila S, Kooli S, Farhat A, Guizani A. Thermal behavior of indirect solar dryer: Nocturnal usage of solar air collector with PCM. Journal of Cleaner Production. 2017;**148**:37-48

[7] Singh D, Mall P, . Experimental investigation of thermal performance of indirect mode solar dryer with phase change material for banana slices. Energy Sources, Part A Recover. Utilization, and Environmental Effects. 2020:1-18. DOI: 10.1080/ 15567036.2020.1810825

[8] Mohana Y, Mohanapriya R, Anukiruthika T, Yoha KS, Moses JA, Anandharamakrishnan C. Solar dryers for food applications: Concepts, designs, and recent advances. Solar Energy. 2020; **208**:321-344

[9] Lingayat AB, Chandramohan VP, Raju VRK, Meda V. A review on indirect type solar dryers for agricultural crops– Dryer setup, its performance, energy storage and important highlights. Applied Energy. 2020;**258**:114005

[10] Jahromi MSB, Kalantar V, Akhijahani HS, Kargarsharifabad H. Recent progress on solar cabinet dryers for agricultural products equipped with energy storage using phase change materials. Journal of Energy Storage. 2022;**51**:104434

[11] Ekechukwu OV, Norton B. Review of solar-energy drying systems II: An overview of solar drying technology. Energy Conversion and Management. 1999;**40**(6):615-655

[12] Basunia MA, Abe T. Thin-layer solar drying characteristics of rough rice under natural convection. Journal of Food Engineering. 2001;**47**(4):295-301

[13] Ghazanfari A, Tabil L Jr, Sokhansanj S. Evaluating a solar dryer for in-shell drying of split pistachio nuts. Drying Technology. 2003;**21**(7): 1357-1368

[14] Seveda MS, Jhajharia D. Design and performance evaluation of solar dryer

for drying of large cardamom (Amomum subulatum). Journal of Renewable and Sustainable Energy. 2012;**4**(6):063129

[15] Sharma A, Chen CR, Lan NV. Solar-energy drying systems: A review. Renewable and Sustainable Energy Reviews. 2009;**13**(6–7):1185-1210

[16] Hii CL, Law CL. Solar drying of major commodity products. In: Solar Drying: Fundamentals, Applications and Innovations. 2012. pp. 73-94

[17] Green MG, Schwarz D. Solar Drying Technology for Food Preservation. Germany: GTZ Publication Eschborn; 2001

[18] Fadhel MI, Abdo RA, Yousif BF, Zaharim A, Sopian K. Thin-layer drying characteristics of banana slices in a force convection indirect solar drying. In Proceedings of the 6th IASME/WSEAS International Conference on Energy and Environment (EE 2011). World Scientific and Engineering Academy and Society Press. February 2011. pp. 310-315

[19] Ghatrehsamani SH, Zomorodian A. Impacts of drying air temperature, bed depth and air flow rate on walnut drying rate in an indirect solar dryer. International Journal of Agriculture Sciences. 2012;**4**(6):253

[20] Atalay H, Cankurtaran E. Energy, exergy, exergoeconomic and exergo-environmental analyses of a large scale solar dryer with PCM energy storage medium. Energy. 2021;**216**:119221

[21] Akbulut A, Durmuş A. Energy and exergy analyses of thin layer drying of mulberry in a forced solar dryer. Energy. 2010;**35**(4):1754-1763

[22] Sami S, Etesami N, Rahimi A. Energy and exergy analysis of an indirect solar cabinet dryer based on mathematical modeling results. Energy. 2011;**36**(5): 2847-2855

[23] Petela R. Exergy of heat radiation. Transactions of the ASME: Journal of Heat Transfer. 1964;**2**:187-192

[24] Tagnamas Z, Kouhila M, Bahammou Y, Lamsyehe H, Moussaoui H, Idlimam A, et al. Drying kinetics and energy analysis of carob seeds (*Ceratonia siliqua* L.) convective solar drying. Journal of Thermal Analysis and Calorimetry. 2022;**147**(3):2281-2291

[25] Fudholi A, Sopian K, Othman MY, Ruslan MH. Energy and exergy analyses of solar drying system of red seaweed. Energy and Buildings. 2014;**68**:121-129

[26] Mugi VR, Chandramohan VP. Energy and exergy analysis of forced and natural convection indirect solar dryers: Estimation of exergy inflow, outflow, losses, exergy efficiencies and sustainability indicators from drying experiments. Journal of Cleaner Production. 2021;**282**:124421

[27] Bhardwaj AK, Kumar R, Kumar S, Goel B, Chauhan R. Energy and exergy analyses of drying medicinal herb in a novel forced convection solar dryer integrated with SHSM and PCM. Sustainable Energy Technologies and Assessments. 2021;**45**:101119

[28] Vijayan S, Arjunan TV, Kumar A. Exergo-environmental analysis of an indirect forced convection solar dryer for drying bitter gourd slices. Renewable Energy. 2020;**146**:2210-2223

[29] Atalay H. Exergoeconomic and environmental impact evaluation of wind energy assisted hybrid solar dryer and conventional solar dryer. Renewable Energy. 2022;**200**:1416-1425

Chapter 2

High Synthetic Image Coding System

Abdallah A. Ibrahim and Loay E. George

Abstract

Compressing an image and reconstructing it without degrading its original quality is one of the challenges that still exist now a day. A coding system that considers both quality and compression rate is implemented in this work. The implemented system applies a high synthetic entropy coding schema to store the compressed image at the smallest size as possible without affecting its original quality. This coding schema is applied with two transform-based techniques, one with Discrete Cosine Transform and the other with Discrete Wavelet Transform. The implemented system was tested with different standard color images and the obtained results with different evaluation metrics have been shown. A comparison was made with some previous related works to test the effectiveness of the implemented coding schema.

Keywords: image coding, image compression, Discrete Cosine Transform (DCT), Discrete Wavelet Transform (DWT), entropy coding, quantization

1. Introduction

In this work, a synthetic image coding schema will be described in detail. The implemented compression system is consisting of four stages to compress the input color image. First, the color transformation resolved from RGB (Red, Green, and Blue) space to YC_bC_r (Y' is the luma component and CB and CR are the blue-difference and red-difference chroma components, respectively). The second stage applies DCT (Discrete Cosine Transform) or DWT (Discrete Wavelet Transform) to produce the transform coefficients for a better representation of image data. These coefficients are quantized using the scalar quantization technique in the third stage. Finally, synthetic entropy coding schema will be implemented to encode the quantized coefficients to produce a stream of bits for the purpose of storing. The entropy encoder work by implementing the adaptive shift coding technique to specify the optimal number of bits needed to store each value of image pixels. The system will be tested on different color images and the results will be shown in detail with various evaluation metrics.

2. The implemented system schema

The system will be implemented with two transform techniques, one with DCT and the other with DWT. In the first schema, the input image is split into blocks of size N × N (where N is multiple of 2), each block is categorized according to its spatial details whether it is a high-frequency block (its pixels are uncorrelated) or low-frequency block that contains correlated pixels by using DPCM (Differential Pulse Coding Modulation) in three different aspects (horizontally, vertically, and diagonally) then the energy of each block is calculated to determine the correlation level between its nearby pixels by using a specified threshold value. Then, image blocks are scanned and converted into 1D vectors using a horizontal scan order. The next step is to apply the 1D-DCT on each vector to produce the transform coefficients. Then, adaptive scalar quantization is applied for both correlated and uncorrelated blocks. The level of the quantization values for each block is different according to the block feature whether it is high-frequency detailed blocks or low-frequency correlated blocks, by taking advantage of that these high-frequency blocks can be treated separately from low frequency to produce better compression. Finally, the entropy encoder is applied to the quantized coefficients to store each coefficient with an optimal number of bits. The encoding and decoding process of DCT are described in detail in our original work in the following paper [1]. **Figure 1** demonstrates the encoder and decoder units.

In the second schema, the image is transformed using wavelet biorthogonal CDF-9/7 (Cohen-Daubechies-Feauveau) to produce the detailed and approximate coefficients. Then, the size of the transformed coefficient will be reduced using progressive scalar quantization. Finally, entropy coding is applied to store these coefficients. DWT encoder and decoder units are presented in **Figure 2**.

3. RGB color space decomposition

RGB color space stands for Red, Green, and Blue, RGB is the most popular color space used, it is device-dependent and it is used for a display system. All colors in the RGB color space are formed using only three-color coordinates: Red, Green, and Blue (RGB) to render the color image. Each color within RGB space ranges from (0 to 255) that is R = {0, 1, 2 ... 255}, G = {0, 1, 2 ... 255} and B = {0, 1, 2 ... 255}. Every pixel within a grayscale image is defined to have a depth of 8 bits per pixel. In other words, the picture consists of different gray values ranging from 0 to 255. In the RGB image, each pixel is a combination of three different channels (Red, Green, and Blue), each of which is considered to be an 8-bit gray-scale image. All other color spaces used in various applications can be extracted from RGB color information [2].

In this work, a true-color image is selected for testing. Each pixel existing in a true-color image comprises of 24-bits. These bits are divided into three parts, each part is set of 8 bits for Red, Green, and Blue channels, respectively. And so, 16,777,216 probable colors could exist within such an image. To process these colors, they are decomposed into their original space channels. These channels hold basic color plane components (Red, Green, and Blue), which hold the color information for each pixel in the image. The values for each channel range from 0 to 255 for each color plane.

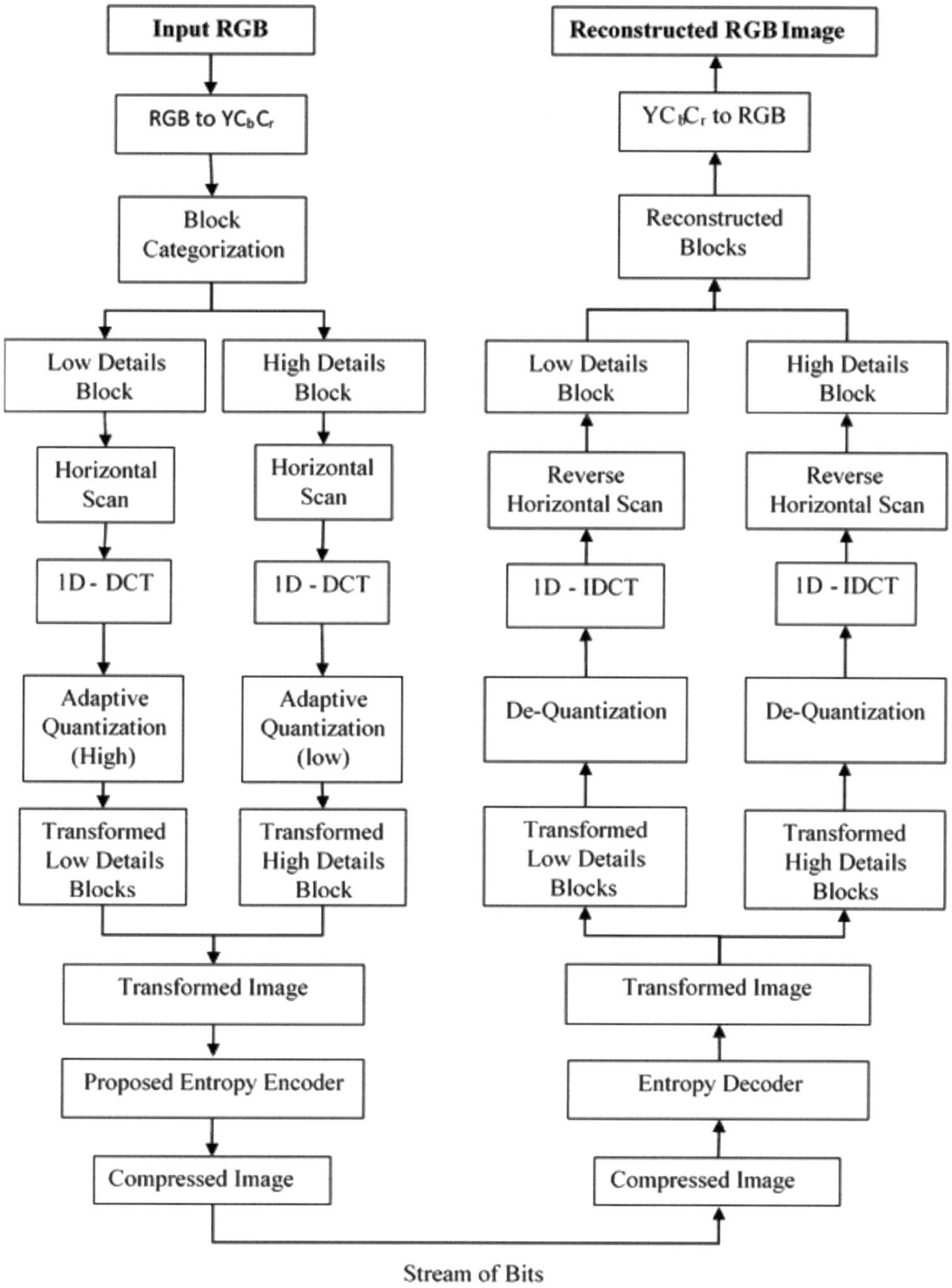

Figure 1.
Block diagram of DCT encoder and decoder [1].

4. RGB to YC_bC_r conversion

YC_bC_r is the most commonly used color space for compression. YC_bC_r is an ascended and offset version of the YUV color space. Y is brightness (luminance or Luma), that is, the gray scale value. U & V are the chrominance or chroma, used to describe the image color and saturation, and used to specify the pixel color. C_b and C_r

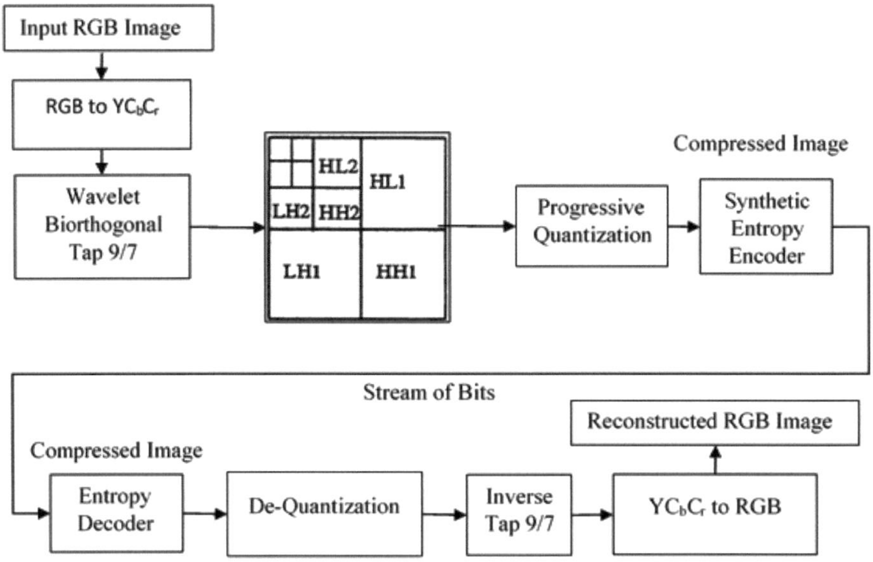

Figure 2.
Block diagram of DWT encoder and decoder.

are the chrominance components, CB is defined as the difference between the blue part of the RGB input signal and the brightness value of the RGB signal, and Cr is the difference between the red part of the RGB input signal and the brightness value of the RGB signal. Since the human visual system (HVS) is more sensitive to changes in luminance than to changes in chrominance, by describing a color in terms of its luminance and chrominance content separately enables more efficient processing and transmission of color signals in many applications, therefore images are compressed more efficiently with this space [3].

The transform from RGB color space to YC_bC_r space is given by the following equation [4]:

$$Y = 16 + \left(\frac{65.738\ \text{Red}}{256}\right) + \left(\frac{129.057\ \text{Green}}{256}\right) + \left(\frac{25.064\ \text{Blue}}{256}\right) \tag{1}$$

$$Cb = 128 + \left(\frac{-37.945\ \text{Red}}{256}\right) - \left(\frac{74.494\ \text{Green}}{256}\right) + \left(\frac{112.439\ \text{Blue}}{256}\right) \tag{2}$$

$$Cr = 128 + \left(\frac{112.439\ \text{Red}}{256}\right) - \left(\frac{94.154\ \text{Green}}{256}\right) - \left(\frac{18.285\ \text{Blue}}{256}\right) \tag{3}$$

where Red, Green, and Blue are the color values of each pixel in the input image multiplied by floating points constants, these constants are scaled by dividing each by $2^8 = 256$.

The inverse transform from YC_bC_r color space to RGB space is given by the following equations:

$$\text{Red} = \left(\frac{298.082\ Y}{256}\right) + \left(\frac{408.583\ Cr}{256}\right) - 222.921 \tag{4}$$

$$\text{Green} = \left(\frac{298.082\ \text{Y}}{256}\right) - \left(\frac{100.291\ \text{Cb}}{256}\right) - \left(\frac{208.120\ \text{Cr}}{256}\right) + 135.576 \tag{5}$$

$$\text{Blue} = \left(\frac{298.082\ \text{Y}}{256}\right) + \left(\frac{516.412\ \text{Cb}}{256}\right) - 276.836 \tag{6}$$

5. CDF-9/7 biorthogonal wavelet transform

To apply the biorthogonal wavelet, transforming the image includes two steps: lifting steps follows by scaling steps. The lifting scheme is implemented through a sequence of phases. It can be identified in three phases: split phase, predict phase, and update phase as shown in **Figure 3**.

In the split phase, the original data is partitioned into two sub-bands: the first sub-band consists of the elements of odd indices and the second consists of the elements of even indices. The predict phase (P) is utilized to get the coefficients of the high-frequency sub-bands details (these coefficients are called wavelet coefficients), and this phase aims to predict the values of the even indices by values of the odd indices. After the predict phase, the update phase (U) is applied to obtain the coefficients of the low-frequency sub-band by making the average value of the output of low-pass coefficients equal to the average values of the original input data. So, the update phase (U) is used to update the even samples by using the previously calculated detail (wavelet) coefficients [6].

6. Progressive scalar quantization

A progressive scalar quantization is applied to quantize the produced wavelet detailed and approximate coefficients. Progressive quantization work by applying quantization in a hierarchal form where each pass in the wavelet transform is quantized with different scales, starting with large quantization scales and decreasing drastically as the number of passes increase.

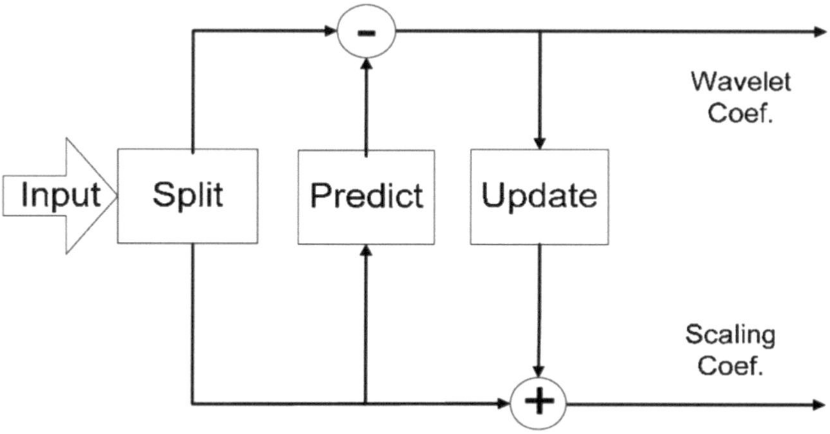

Figure 3.
Lifting scheme transform [5].

The quantization step used to quantize the coefficients of each sub-band calculated, according to the following eq. [7]:

$$Q_s = \begin{cases} Q0 \text{ for LL band in the } n^{th} \text{ Level} \\ Q\alpha^{NPass-1} \text{ for LH.HL in } n^{th} \text{ Level} \\ Q\beta\alpha^{NPass-1} \text{for HH in } n^{th} \text{ Level} \end{cases} \quad (7)$$

where N_{Pass} is the wavelet level number (i.e., the pass number), α is the descending rate parameter, and ***β*** is the Beta multiplication (such that Q0 = 2, $Q \geq 1$, $\alpha \leq 1$, $\beta \geq 1$). The value of the quantization step is decreased using linear, progressive relationship, and its value for HH sub-band is greater than its value for the corresponding HL and LH sub-bands because it is multiplied by β (≥ 1).

7. Synthetic entropy encoder

Synthetic entropy coding is applied to represent the quantized coefficients as a sequence of 0 and 1, by storing the optimal number of bits needed for each coefficient. The flowchart of the implemented synthetic entropy encoder is shown in **Figure 4**.

Firstly, the input array of coefficients is partitioned as a block of size (N*N), where N is the size of the block used in DCT divided by 2 or the size of the LL band used in DWT. Then, the block with all zero coefficients will be discarded and the block with nonzero values will be converted into 1-D vector.

The next step is to map each negative value into a positive one to get rid of the negative sign and also avoid the coding complexity when storing these numbers. This is simply done by mapping all negative values to positive odd numbers while the positive values are mapped to be even numbers. The mapped numbers are produced by applying the following equation [7]:

$$\text{Pi} = \begin{cases} 2\,Pi \text{ } if \text{ } Pi \geq 0 \\ (/Pi/ * 2) - 1 \text{ } if \text{ } Pi < 0 \end{cases} \quad (8)$$

where Pi is the coefficient value in the incoming sequence.

The inverse mapping process is implemented using the following equation:

$$\text{P'i} = \begin{cases} Pi/2 \text{ } if \text{ } Pi \text{ } is \text{ } even \text{ } number \\ (-Pi + 1)/2 \text{ } if \text{ } Pi \text{ } is \text{ } odd \text{ } number \end{cases} \quad (9)$$

The next step is to decompose each vector by moving each nonzero value (> = 1) into a new vector and replacing its original location with one. Then, the run length encoding process is applied to the original vector [1].

The resultant vectors from the previous stage (Nonzero Vector and Run Vector) are further separated into two sub-vectors, according to the mean value. The first vector will contain the values less than the mean value and the second vector will contain the values that are greater than or equal to the mean value and then subtract the second vector from the mean value to reduce its scaling range.

Finally, the adaptive shift coding technique is applied to assign the optimal number of bits required to store each coefficient [1].

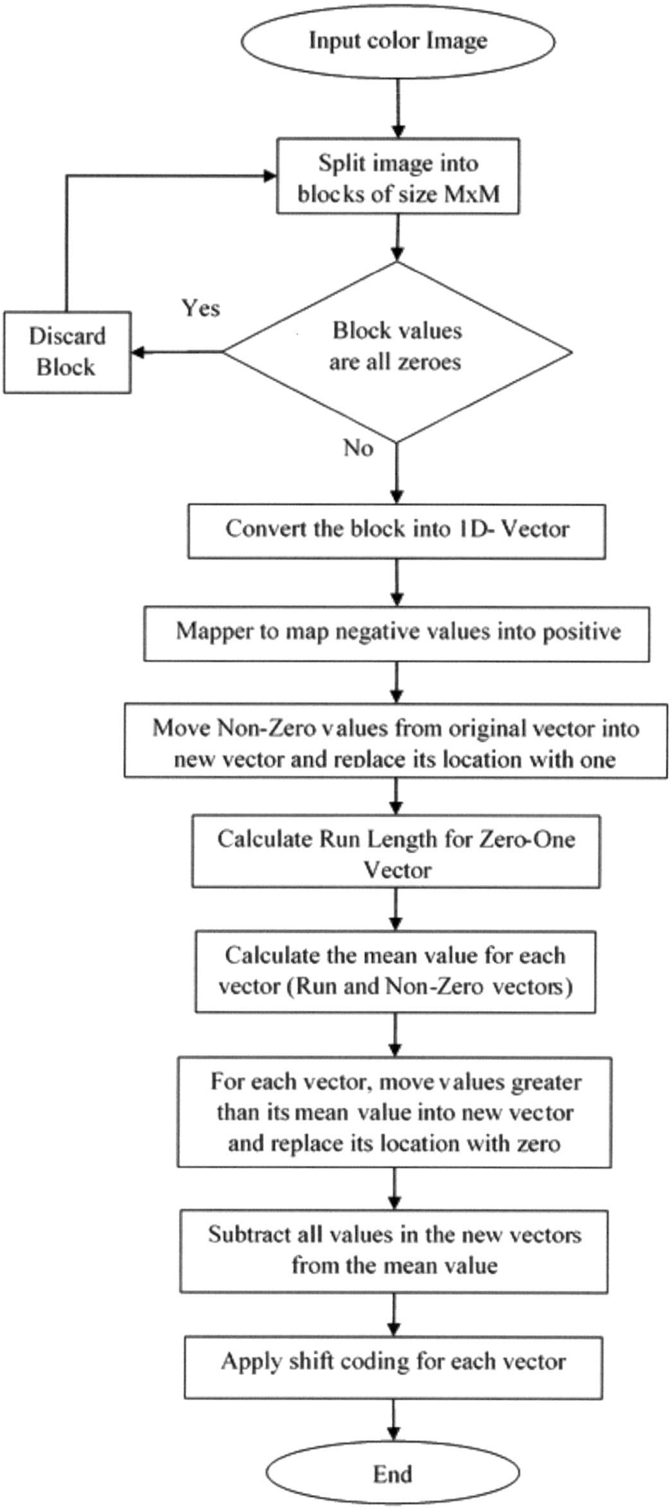

Figure 4.
The implemented synthetic entropy encoder flowchart.

8. Evaluation metrics

A set of evaluation metrics have been taken to investigate the performance of the compression method applied.

Compression ratio (CR) is defined as the ratio of the size between the original image and the compressed image size. They can be measured in (bits, bytes, kilobytes, etc.). The higher CR is the better compression technique used to compress the image [8].

$$CR = \frac{\text{Original file Size}}{\text{Compressed}} \tag{10}$$

Mean Square Error (MSE) is computed by averaging the cumulative squared intensity error differences between the original image and the reconstructed image. MSE value closest to zero is better, and the reconstructed image quality is poor when MSE is large, thus MSE must be as low as possible for effective compression. MSE is calculated through the following equation [9]:

$$MSE = \frac{1}{m\,n}\sum_{i=0}^{m-1}\sum_{j=0}^{n-1}[\,X(i.j) - Y(i.j)]^2 \tag{11}$$

where X is the original image pixel and Y is the compressed image. The dimension of the images is $m \times n$.

Peak-Signal-to-Noise Ratio (PSNR) is the ratio between maximum signal power, which is considered as the original image and the power of distorting noise obtained from MSE. The higher the PSNR, the better the quality of the reconstructed image. Typical values for lossy compression of an image are between 28 dB and 40 dB. The PSNR can be given by the following equation [10]:

$$PSNR = 10\ \log_{10}\left(\frac{MAX_x^2}{MSE}\right) \tag{12}$$

where MAX^2 is the power of maximum intensity value in the original image X.

Bit Per Pixel (BPP) is defined as a number of bits required to store each pixel in an image. Pixels must be coded efficiently to reduce redundancy, hence reducing storage requirements. BPP is calculated using the following equation [11]:

$$BPP = \frac{B}{MxN} \tag{13}$$

$$\text{or } BPP = \frac{24\text{bits}}{CR} \tag{14}$$

where B is a number of bits after compression and $M \times N$ is the total number of pixels in an image.

Compression gain (CG) is defined as the amount of compression gained after the image is compressed. The CG of a digital image is calculated by the following equations:

$$CG = \frac{\text{original size} - \text{compressed size}}{\text{original size}}\ x\ 100\% \tag{15}$$

$$\text{or} \qquad CG = 1 - \frac{1}{CR} \times 100\% \tag{16}$$

Structural similarity index measure (SSIM) aims to measure quality by capturing the similarity of images. Three aspects of similarity: luminance, contrast, and structure are determined and their product is measured [9].

9. Experimental results

This section is intended to examine the results of the proposed system in detail for both DCT and DWT modules. The detailed results of many experiment tests are presented and discussed to evaluate the performance behavior of the established system. The adopted system and all additional programs for the testing purpose have been established using visual studio (C sharp programming language).

A set of standard true-color images has been utilized to test and evaluate the system's performance. These images are of type "Bitmap" format in which each pixel is stored in 24-bit true color. The image files "Lena.bmp," "Barbara.bmp," "Peppers.bmp," (smoothed images), and "Baboon.bmp" (sharp edge image). All test images are of size 256 × 256 for width and height. **Figure 5** presents these images.

A set of results tables is presented for evaluation. Each test image is evaluated using two modules. One shows the results of compression when the DCT technique is applied and the second module shows the results of compression when the DWT technique is applied. Each table shows the effectiveness of each control parameter used in the system on the resulting compressed image. The test results are evaluated and compared based on fidelity criteria measurements (i.e., MSE, PSNR, SSIM, CR, CG, and BPP). The real time of encoding and decoding process is also presented.

The effect of the following control parameters has been investigated to test the results of the proposed system using the DCT technique:

BS: is the block size used in DCT.

Thr: is the threshold value used to determine the importance of each block.

Q0, Q1, and α: are the quantization control parameters used to calculate the quantization step value.

The effects of each parameter are explored by varying the value of each parameter from minimum value to maximum. **Table 1** represents the assumed default range of the considered control parameters used with the DCT module.

Figure 5.
Test color images.

Parameter	Range
BS	[8, 16]
Q0	[32,128]
Q1	[1–10]
A	[0.1–1]
Thr	[1–20]

Table 1.
The range of the control parameters for the system when using DCT.

The parameters used to control the level of compression by using the DWT technique are instigated as the following:

N.Pass: the number of passes used in the wavelet transform.

Q0: used to quantize the approximation LL band.

Q1: used to quantize the details sub-bands (LH, HL, and HH).

α: the descending rate parameter used to decrease the quantization of details bands as the number of passes increase.

β: Beta multiplication parameter used to scale the quantization of HH sub-band.

The range of these parameters after making a comprehensive set of tests is shown in **Table 2**.

A set of results tables in our previous work [1] presents the system performance when applying the DCT module in terms of MSE, PSNR, SSIM, CR, CG, and BPP for Lena, Barbara, Peppers, and Baboon images, respectively. It shows that the fidelity level in terms of PSNR and SSIM increased while the CR and CG decreased.

In the same manner, the system performance was evaluated when applying DWT module on the same color images with the effects of control parameters shown in **Table 2**, **Tables 3–6** present these results.

Parameter	Range
N.Pass	[3–7]
Q0	[2]
Q1	[1–35]
α	[0.3–1]
β	[1.2]

Table 2.
The default range of the control parameters for the system when using DWT.

MSE	PSNR (dB)	SSIM	CR	CG (%)	BPP	Time (s)	
						ET	DT
109.437	27.739	0.851	55.492	98.198	0.432	0.057	0.045
98.590	28.192	0.851	54.538	98.166	0.440	0.096	0.062
81.313	29.029	0.871	45.776	97.815	0.524	0.073	0.052
64.585	30.029	0.895	35.165	97.156	0.682	0.056	0.051

MSE	PSNR (dB)	SSIM	CR	CG (%)	BPP	Time (s)	
						ET	DT
51.441	31.018	0.913	29.067	96.560	0.826	0.060	0.047
40.878	32.016	0.926	23.431	95.732	1.024	0.071	0.054
32.325	33.035	0.938	19.528	94.879	1.229	0.081	0.054
24.978	34.155	0.952	16.289	93.861	1.473	0.063	0.062
20.527	35.008	0.960	13.347	92.508	1.798	0.066	0.060
16.302	36.008	0.965	10.396	90.381	2.308	0.082	0.074
12.846	37.043	0.974	8.794	88.628	2.729	0.077	0.054
9.987	38.136	0.980	7.005	85.724	3.426	0.104	0.090
8.124	39.033	0.983	5.529	81.915	4.340	0.102	0.079
6.307	40.132	0.987	4.684	78.650	5.124	0.106	0.072

Table 3.
Test results after applying compression with DWT module on color "Lena" test image.

MSE	PSNR (dB)	SSIM	CR	CG (%)	BPP	Time (s)	
						ET	DT
97.922	28.222	0.851	52.373	98.091	0.458	0.067	0.054
79.927	29.104	0.872	43.905	97.722	0.547	0.065	0.056
64.555	30.032	0.887	36.908	97.291	0.650	0.088	0.110
50.343	31.111	0.909	30.458	96.717	0.788	0.075	0.061
40.511	32.055	0.923	24.963	95.994	0.961	0.070	0.050
32.338	33.034	0.937	21.911	95.436	1.095	0.062	0.061
25.745	34.024	0.947	18.661	94.641	1.286	0.064	0.059
19.256	35.285	0.958	15.815	93.677	1.518	0.062	0.066
15.426	36.248	0.968	12.914	92.257	1.858	0.074	0.075
12.480	37.169	0.974	11.334	91.177	2.118	0.073	0.064
9.482	38.362	0.980	9.652	89.640	2.486	0.080	0.054
7.683	39.275	0.984	7.882	87.313	3.045	0.082	0.083
6.185	40.217	0.986	6.811	85.317	3.524	0.081	0.058

Table 4.
Test results after applying compression with DWT module on color "Barbara" test image.

MSE	PSNR (dB)	SSIM	CR	CG (%)	BPP	Time (s)	
						ET	DT
111.022	27.677	0.852	30.219	96.691	0.794	0.065	0.052
100.160	28.124	0.862	26.848	96.275	0.894	0.094	0.080
81.561	29.016	0.877	21.840	95.421	1.099	0.085	0.063
77.301	29.249	0.881	21.734	95.399	1.104	0.084	0.067

MSE	PSNR (dB)	SSIM	CR	CG (%)	BPP	Time (s)	
						ET	DT
63.036	30.135	0.897	17.096	94.151	1.404	0.072	0.053
51.382	31.023	0.910	13.464	92.573	1.782	0.064	0.054
40.738	32.031	0.924	9.420	89.384	2.548	0.072	0.059
30.476	33.291	0.945	6.819	85.335	3.520	0.088	0.064
25.562	34.055	0.954	5.687	82.415	4.220	0.075	0.066
19.684	35.190	0.966	4.143	75.863	5.793	0.114	0.090
15.180	36.318	0.975	3.667	72.733	6.544	0.120	0.111
7.750	39.238	0.988	2.884	65.328	8.321	0.098	0.086
8.313	38.933	0.988	2.880	65.284	8.332	0.122	0.097
4.468	41.630	0.994	2.095	52.269	11.455	0.082	0.081

Table 5.
Test results after applying compression with DWT module on color "Peppers" test image.

MSE	PSNR (dB)	SSIM	CR	CG (%)	BPP	Time (s)	
						ET	DT
241.207	24.307	0.856	11.238	91.102	2.136	0.097	0.080
199.244	25.137	0.880	9.145	89.065	2.625	0.077	0.065
153.916	26.258	0.913	6.902	85.512	3.477	0.085	0.073
128.413	27.045	0.930	5.797	82.749	4.140	0.080	0.065
97.608	28.236	0.946	4.824	79.270	4.975	0.114	0.064
78.884	29.161	0.958	3.997	74.981	6.005	0.157	0.135
63.264	30.119	0.966	3.613	72.324	6.642	0.130	0.115
49.856	31.154	0.973	3.242	69.155	7.403	0.127	0.112
40.391	32.068	0.978	3.049	67.207	7.870	0.100	0.079
27.843	33.684	0.985	2.719	63.227	8.826	0.108	0.079
25.160	34.124	0.986	2.392	58.193	10.034	0.103	0.070
18.913	35.363	0.990	2.384	58.053	10.067	0.158	0.097
15.876	36.123	0.992	2.098	52.336	11.439	0.114	0.077
9.655	38.283	0.996	1.986	49.651	12.084	0.106	0.076
5.201	40.970	0.998	1.553	35.617	15.452	0.138	0.097
4.759	41.355	0.998	1.523	34.324	15.762	0.087	0.073

Table 6.
Test results after applying compression with DWT module on color "Baboon" test image.

Figures 6–9 show the tradeoff between CR and PSNR when applying compression on the test images (Lena, Barbara, Peppers, and Baboon), respectively.

A set of reconstructed color images is shown in **Figure 10** with different fidelity levels, where PSNR, SSIM, CR, and BPP values are varied.

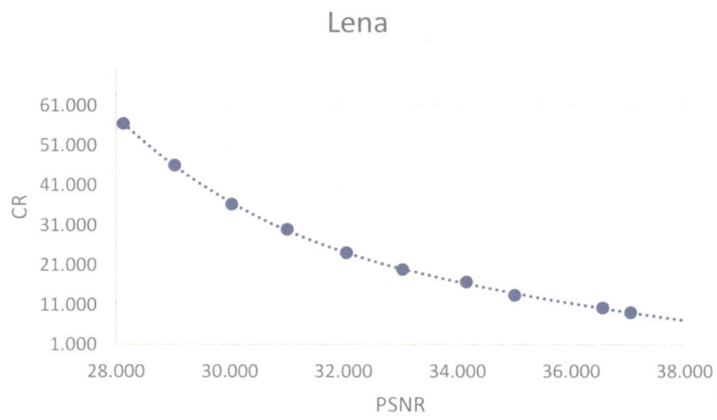

Figure 6.
A tradeoff between CR and PSNR for color Lena image test results.

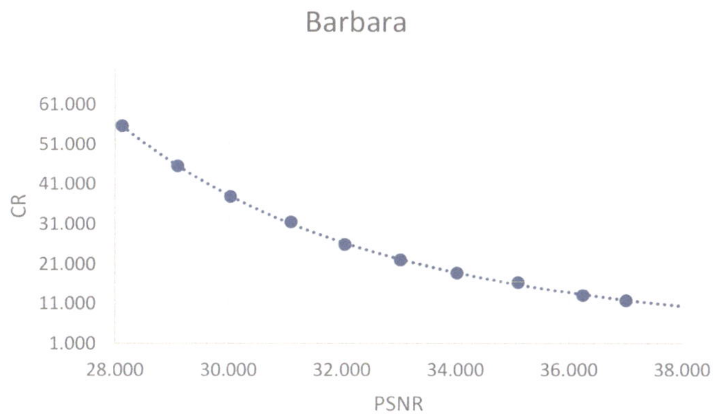

Figure 7.
A tradeoff between CR and PSNR for color Barbara image test results.

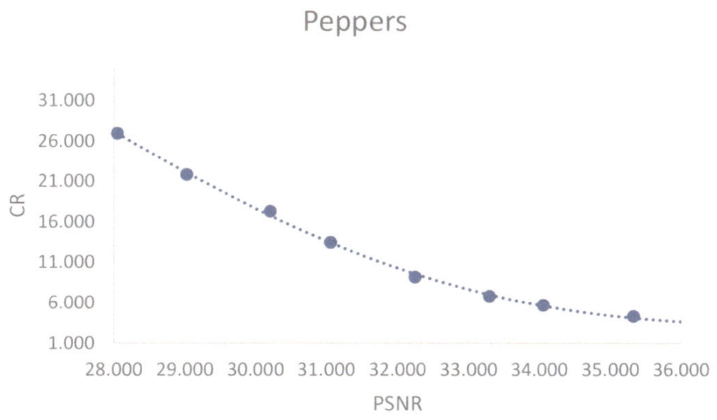

Figure 8.
A tradeoff between CR and PSNR for color Peppers image test results.

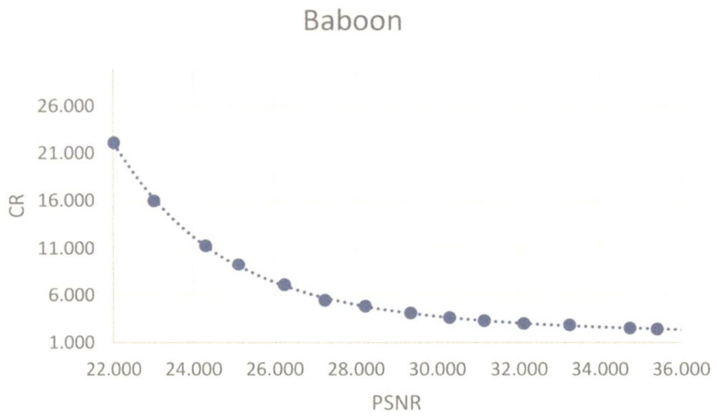

Figure 9.
A tradeoff between CR and PSNR for color Baboon image test results.

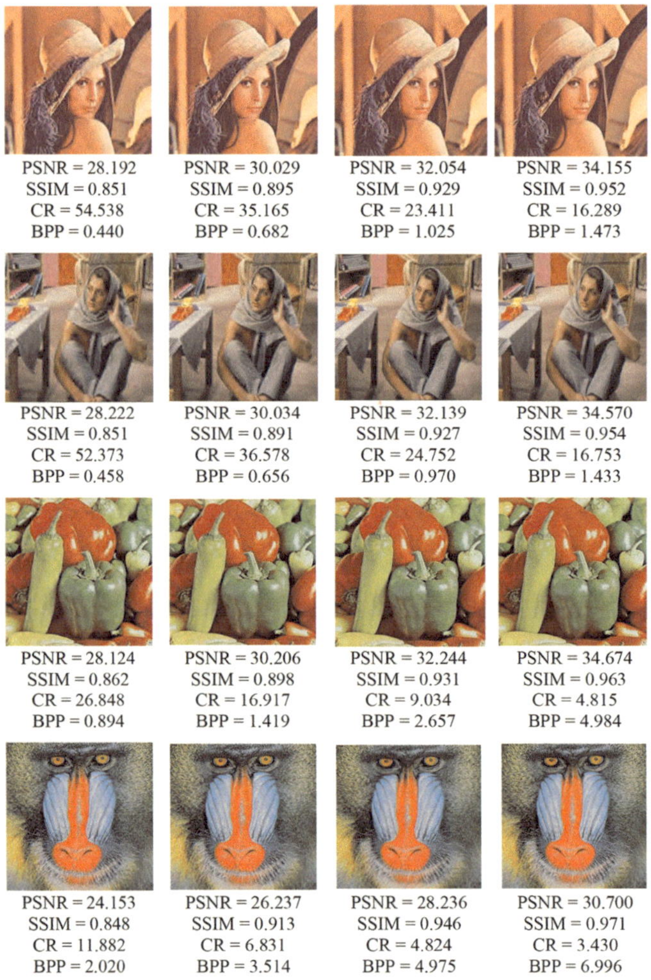

Figure 10.
Samples of the reconstructed color images where the PSNR, SSIM, CR, and BPP values are varied.

10. Comparisons with previous works

In this section, the results of the implemented system have been compared with some previously published methods and also with the standard JPEG system.

Table 7 lists the compression results attained by the proposed system with those given in previous studies in terms of (CR, BPP, PSNR, and SSIM), taking into consideration that in these studies same images have been used. The listed results demonstrate that the proposed system outperforms other methods.

11. Conclusions

In this chapter, a high entropy image coding system was implemented, and the system shows remarkable results compared to the existing approaches. A high compression ratio was obtained while maintaining image quality without distortion. The proposed entropy encoder has a positive significant impact on the results. It is important that the transformed image coefficients must be processed in a way to be suitable for the coding process in order to store it with a minimum number of bits.

Test Image	Reference	Size	CR	BPP	PSNR	SSIM
Color Lena	[12]	512 × 512	—	1.0	33.11	0.9583
	[13]	512 × 512	12.67	—	33.86	—
	[14]	512 × 512	—	0.75	33.01	—
	[15]	512 × 512	—	0.4	30.83	—
	[16]	512 × 512	—	0.82	33.15	—
	[11]	512 × 512	—	0.73	31.556	—
	JPEG standard	512 × 512	32.650	0.980	30.05	0.798
	Proposed	512 × 512	39.678	0.6	33.134	0.963
Color peppers	[14]	512 × 512	—	0.64	30.49	—
	[15]	512 × 512	—	0.4	28.53	—
	[16]	512 × 512	—	0.95	30.97	—
	[11]	512 × 512	—	0.9	31.604	—
	JPEG standard	512 × 512	29.528	0.974	28.588	0.839
	Proposed	512 × 512	33.555	0.7	29.781	0.955
Color baboon	[12]	512 × 512	—	1.0	29.74	0.6656
	[13]	512 × 512	6.69	—	30.15	—
	[11]	512 × 512	—	0.8	29.792	—
	JPEG standard	512 × 512	24.403	0.978	15.332	1.565
	Proposed	512 × 512	24.672	0.941	15.971	1.5

Table 7.
Comparison between the implemented system results and some related works used to encode different standard images.

From the obtained test results, the proposed system shows outperformed results when dealing with smooth images like (Lena, Barbara, and Peppers) compared to the existing approaches, but it has a major drawback in a sharp image like Baboon.

As a summary, the best-attained compression gains for the test images (Lena, Barbara, Peppers, and Baboon) are 98.16%, 98.09%, 96.28%, and 91.10%, respectively, where PSNR values are in an intermediate range.

Author details

Abdallah A. Ibrahim[1*] and Loay E. George[2]

1 University of Baghdad, Baghdad, Iraq

2 University of Information Technology and Communications, Baghdad, Iraq

*Address all correspondence to: abdullah.i@sc.uobaghdad.edu.iq

References

[1] Ibrahim AA, George LE, Hassan EK. Color image compression system by using block categorization based on spatial details and DCT followed by improved entropy encoder. Iraqi Journal of Science. 2020;**61**(11):3127-3140

[2] Pm N, Chezian R. Various color spaces and color space conversion algorithms. Journal of Global Research in Computer Science. 2013;**2013**:44-48

[3] Szedo G. Color-Space Converter: RGB to YCrCb2007. pp. 1-16. Available from: https://www.researchgate.net/publication/277048646_R_Color-Space_Converter_RGB_to_YCrCb

[4] ITU-R. Studio Encoding Parameters of Digital Television for Standard 4:3 and Wide-screen 16:9 Aspect Ratios. International Telecommunication Union; 2011. Available from: https://www.itu.int/rec/R-REC-BT.601/

[5] Shirvaikar M, Bushnaq T. VHDL implementation of wavelet packet transforms using SIMULINK tools. Real-Time Image Processing. 2008;**2008**:6811

[6] Lian C-J, Chen K-F, Chen H-H, Chen L-G. Lifting based discrete wavelet transform architecture for JPEG2000. In: The 2001 IEEE International Symposium on Circuits and Systems (Cat. No.01CH37196). 2001. pp. 445-448. Available from: https://www.semanticscholar.org/paper/Lifting-based-discrete-wavelet-transform-for-Lian-Chen/cc46ac3f1f89ec38b8b641dd43d8254b8c8637bb

[7] Ahmed SD, George LE, Dhannoon BN. The use of cubic bezier interpolation, biorthogonal wavelet and quadtree coding to compress color images. British Journal of Applied Science & Technology. 2015;**11**(4):1-11

[8] Anandan P, Sabeenian RS. Medical image compression using wrapping based fast discrete curvelet transform and arithmetic coding. Circuits and Systems. 2016;**2016**:2059-2069

[9] Varnan C, Jagan A, Kaur J, Jyoti D, Rao DS. Image quality assessment techniques pn spatial domain. International Journal of Computer Science and Technology. 2011;**2**(3):177-184

[10] Silpa K, Mastani S. Comparison of image quality metrics. International Journal of Engineering Research & Technology (IJERT). 2012;**1**(4):1-5

[11] Nigam AK, Khare P, Srivastava VK. Image compression using hybrid approach and adaptive scanning for color images. Advances in VLSI, Communication, and Signal Processing. 2020;**587**:761-774

[12] Kalavathi P, Boopathiraja S. A wavelet based image compression with RLC encoder. In: Computational Methods, Communication Techniques and Informatics. pp. 289-292. Available from: https://www.researchgate.net/profile/Kalaiselvi-Thiruvenkadam/publication/319077453_Computation_Methods_Communication_Techniques_and_Informatics/links/5993f440aca272ec9087ac00/Computation-Methods-Communication-Techniques-and-Informatics.pdf#page=307

[13] Hashim AT, Ali SA. Color image compression using DPCM with DCT, DWT and quadtree coding scheme. Engineering and Technology Journal. 2020;**34**:585-597

[14] Messaoudi A, Benchabane F, Srairi K. DCT-based Color Image Compression Algorithm using Adaptive

Block Scanning, Signal, Image and Video Processing. Springer; 2019. Available from: https://link.springer.com/article/10.1007/s11760-019-01492-7

[15] Zhou Y, Wang C, Zhou X. DCT-based color image compression algorithm using an efficient lossless encoder. Image and Video Processing. 2018;**2018**:450-454

[16] Boucetta A, Melkemi KE. DWT Based-Approach for Color Image Compression Using Genetic Algorithm. 2012. pp. 476-484. Available from: https://books.google.iq/books?hl=en&lr=&id=mwC7BQAAQBAJ&oi=fnd&pg=PA476&dq=DWT+Based-Approach+for+Color+Image+Compression+Using+Genetic+Algorithm&ots=n5iiuuQd9-&sig=ToXcTrsRUaTtnSwckkq7to2XVZY&redir_esc=y#v=onepage&q=DWT%20Based-Approach%20for%20Color%20Image%20Compression%20Using%20Genetic%20Algorithm&f=false

Chapter 3

Nonreciprocal Photovoltaics: The Path to Conversion of Entire Power-Beam Exergy

Andrei Sergeev and Kimberly Sablon

Abstract

Nonreciprocal photonic management can shift the absorption-emission balance in favor of absorption and enhance the conversion efficiency beyond the detailed balance Shockley - Queisser limit. Nonreciprocal photovoltaic (PV) cells can provide the conversion of the entire exergy (Helmholtz free energy) of quasi-monochromatic radiation into electric power. Recent discoveries in electromagnetics have demonstrated the ability to break Kirchhoff's reciprocity in a variety of ways. The absorption-emission nonreciprocity may be realized via dissipationless one-way optical components as well as via the greenhouse-type electron-photon kinetics that traps the low-energy near-bandgap photons in the cell. We calculate the limiting performance of the nonreciprocal dissipationless monochromatic converter and discuss the limiting efficiency of the nonreciprocal converter based on the greenhouse effect. We also perform detailed modeling of the greenhouse effect in the GaAs PV converter and determined its PV performance for conversion of 809 nm laser radiation. In perovskite PV cells the greenhouse filter establishes a sharp absorption edge and reduces conversion losses related to the distributed PV bandgap and laser-cell matching losses.

Keywords: photovoltaic conversion, photon exergy, absorption-emission no reciprocity, power beaming, greenhouse effect

1. Introduction

Power delivery by a laser beam is an emerging technology with numerous potential applications. The capabilities of unmanned aerial vehicles, various robotic platforms, and sensor networks will be strongly enhanced due to remote charging. Currently all technological components for power beam delivery and conversion are commercially available. Integration of lasers with photovoltaic converters requires matching laser quanta to semiconductor material characteristics. In traditional design, the optimal bandgap value is determined by a tradeoff between the low near-bandgap absorption and thermalization losses [1, 2]. The optimal bandgap depends on laser power, optoelectronic properties of semiconductor material, and cell design [1]. Usually, the optimal bandgap wavelength exceeds the semiconductor bandgap by 20–80 nm. Even in optimized converters, the photoelectron thermalization and weak near-bandgap absorption produce notable losses [2]. Greenhouse-type filter that traps photons with wavelengths above the laser wavelength is a tool of choice to eliminate the laser-cell matching losses [3, 4]. Moreover, as it is shown in this work, the greenhouse filter can generate the greenhouse effect in the cell, which strongly reduces the emission from the cell and, in this way, increases the conversion efficiency beyond the detailed-balance Shockley - Queisser limit. The greenhouse PV effect is a special case of nonreciprocal photonic management, which violates Kirchhoff's reciprocity between absorption and emission [5, 6].

The atmospheric greenhouse effect was discovered by Joseph Fourier, who calculated the balance between the incoming solar power and the outgoing power of emitted radiation and found the Earth's average temperature near of 0°F To shift the absorption-emission balance in a favor of absorption and get the average temperature of $\sim$ 60°F, Fourier proposed that the emitted radiation is trapped by the atmosphere similar to the way a greenhouse glass traps the heat [7]. The greenhouse glass transmits high frequency (high energy) radiation, which is absorbed by the greenhouse media and plants. The media convert the radiation into heat and emit thermal low frequency (low energy) radiation, which is reflected back to the greenhouse by the greenhouse glass. Plants use all high and low energy radiation and convert the radiation into biochemical energy. Photovoltaic conversion directly transfers radiation into electric power via the generation of electrons and holes, which lose part of their energy to crystalline lattice and are accumulated near the bandgap edge. Photocarriers can recombine via nonradiative processes, which convert solar energy into heat. Photocarriers can also recombine via the radiative process and emit bandgap photons. In the absence of nonradiative processes, the photocarrier density and PV performance are determined by the detailed balance between the absorbed and emitted radiation fluxes. As was shown by Shockley and Queisser (SQ), this absorption-emission balance defines a fundamental limit of the photovoltaic conversion efficiency, which in traditional cell design depends on a single material parameter, – the semiconductor bandgap [8].

Recently proposed greenhouse design [3] mimics the greenhouse effect and nonequilibrium greenhouse processes, such as trapping of near bandgap radiation and reusing it for conversion into electricity. Greenhouse filter is placed at the front surface of a cell and the back surface mirror. It establishes a photonic bandgap above the semiconductor bandgap and traps the photons with energies below the photonic bandgap (see **Figure 1**). The mirror may be a wideband Bragg reflector, or photonic crystal reflector, or metallic reflector with small absorption. In the greenhouse design, the photon emission from the converter is limited by recombination processes of hot photo carriers that emit photons with energy above the photonic bandgap. Bandgap photons emitted by the near-bandgap photocarriers are recycled by the filter and reused in the cell for PV conversion. In the PV greenhouse effect, the nonradiative recombination plays a role of a greenhouse media, which heats the greenhouse. The radiative processes play a role in photosynthesis, which convert solar energy into biochemical energy used by plants. Therefore, the greenhouse PV design requires high-quality PV materials with weak nonradiative recombination, i.e. materials with high internal quantum (radiative) efficiency. Fast progress in traditional semiconductor materials and the development of novel optoelectronic materials raises principle questions about photonic management for PV conversion.

Can the greenhouse filter increase the solar light conversion efficiency beyond the SQ limit? The answer substantially depends on a form of the nonequilibrium distribution function of photo-generated carriers. In the greenhouse converter, we have

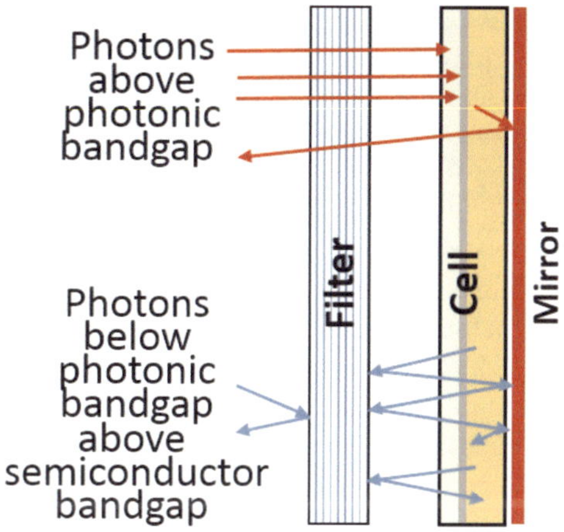

Figure 1.
PV converter with the greenhouse filter, which establishes the photonic bandgap above the semiconductor bandgap and traps photons with energy below the photonic bandgap inside the converter.

two characteristic bandgaps. The accumulation and collection of photocarriers occur near the semiconductor bandgap. The photonic bandgap controls the absorption-emission balance in the cell. In this design, only photons emitted by hot electrons can leave the converter. The population of hot electrons and emission from the cell are controlled by the cooling of the hot electrons. If photocarriers near the semiconductor bandgap and hot energy photocarriers above the photonic bandgap have the same photo-induced chemical potential, the greenhouse filter just shifts the PV bandgap from the semiconductor bandgap to the photonic bandgap. In other words, in this quasi-equilibrium case (actually, in a chemical equilibrium between low energy and high energy photocarriers), the limiting conversion efficiency is the SQ efficiency with the PV bandgap equaled to the photonic bandgap. However, hot photocarriers usually rapidly lose their energy and, as a result, the density of hot photocarriers decreases, and their chemical potential is significantly reduced in comparison with the chemical potential of carriers near the semiconductor bandgap. In this nonequilibrium regime with fast photocarrier cooling, the conversion efficiency may exceed the SQ limit. To realize the nonequilibrium regime, the difference between photonic and semiconductor bandgaps should substantially exceed the thermal energy. Therefore, the photonic bandgap should at least be $2 - 3\ k_B T =$ 50-75 meV higher than the semiconductor bandgap.

Let us highlight that in the traditional cell design all conversion processes occur in the narrow energy interval above the semiconductor bandgap. Therefore, the conversion efficiency turns out to be insensitive to details of electron, photon, and phonon processes. In particular, the detailed modeling of PV conversion as a function of characteristic photoelectron relaxation and extraction times has shown that optimization of traditional design and operating regimes does not allow for surpassing the SQ limit [9]. The greenhouse design provides a splitting of key conversion processes. The photocarrier accumulation and collection take place near the semiconductor bandgap, while photocarrier recombination with photon escape occurs above the photonic bandgap. The filter provides an effective tool to reduce cell emission and shift the absorption-emission balance in favor of absorption.

To avoid cell heating, the greenhouse design requires photovoltaic materials with low nonradiative losses, i.e. high quantum efficiencies. Among traditional photovoltaic materials, GaAs has the highest radiative efficiency. In particular, high-quality GaAs with internal quantum efficiency (also termed internal radiative efficiency) of 99.7% provides solar cells with external radiative efficiency (ERE) above 30% [10–12]. In silicon, the Auger recombination is stronger than radiative recombination and the corresponding ERE of 1% substantially limits the conversion efficiency. Recent progress in emerging photovoltaic materials, – organic materials, dye-sensitized, CuInGaSe (CIGS), and lead-halide perovskites, – has demonstrated strong improvements in ERE. In particular, the ERE of CIGS currently exceeds 24%. ERE is the integral characteristic, which may be calculated via the special averaging of external quantum efficiency (EQE) in the relatively narrow spectral range around the absorption threshold, which in traditional semiconductors coincides with the bandgap [13]. The perovskite materials demonstrate EQE values very close to unity in wide spectral ranges and gradual reduction of EQE near threshold energy [14, 15], which substantially limits the conversion efficiency of traditional PV cells.

In this work, we investigate and optimize greenhouse photonic management for photovoltaic conversion of monochromatic radiation. Recent progress in electromagnetics has demonstrated the ability to break the absorption-emission reciprocity in a variety of ways. To understand the advantages and limitations of greenhouse photonic management, in Section 2 we derive the monochromatic detailed-balance efficiency and in Section 3 we consider the limiting nonreciprocal monochromatic efficiency realized via dissipationless nonreciprocal optical components. In Section 4, we investigate the greenhouse photonic management and present results of simulations of conversion of 10 W/cm^2 laser radiation with a wavelength around 809 nm by GaAs PV cell with the greenhouse filter. In p-doped A_3B_5 semiconductors the electron cooling is realized due to energy transfer from hot electrons to holes [16]. We determine the corresponding non-equilibrium distribution with reduced chemical potential of hot photocarriers and calculate the cell performance. Finally, we briefly discuss enhanced PV conversion in perovskite cells due to greenhouse filtering, which establishes a sharp absorption edge.

2. Detailed-balance limiting efficiency for monochromatic radiation

Let us start with the SQ detailed balance approach, which is based on two assumptions. The first assumption is the reciprocity of photonic (radiative) processes. In the equilibrium, the emitted radiation is exactly given by the absorbed radiation reversed in time. In other words, following Kirchhoff's law, emissivity, $e(\lambda, \mathbf{n})$, and absorptivity, $\alpha(\lambda, -\mathbf{n})$, are equal for any photon wavelength and any propagation direction, $\mathbf{n}$. Second, in PV conversion, photocarriers and emitted photons reach chemical equilibrium. Thus, photocarriers and photons have the same light-induced chemical potential. The photocarriers are collected in the narrow energy range above the semiconductor bandgap and photons are also emitted in the same range. Therefore, assumption about the chemical equilibrium between photocarriers and photons is only essential for this narrow range, which is of the order of thermal energy $k_B T_0$..

The generalized SQ model is described by three parameters. The detailed balance is taken into account by the ratio of the absorbed flux to the equilibrium emitted flux, N_{ab}/N_{em}, where N_{ab} is the absorbed flux, N_{em} is emitted photonic flux in the equilibrium at the device operating temperature T_0. The incoming flux and its absorption by the cell are characterized by the average photon energy in the absorbed flux, ϵ^*. Nonradiative recombination losses and losses in the photocarrier collection are described by the external radiative efficiency of the cell, k_{ERE}. The generalized SQ model allows for a rather simple analytical solution [17]. The SQ open-circuit voltage and conversion efficiency are given by

$$V_{OC} = \frac{k_B T_0}{q} \cdot \ln A_k, \qquad A_k = k_{ERE} \cdot \frac{N_{ab}}{N_{em}} \tag{1}$$

$$\eta_{SQ} = \frac{k_B T_0}{\epsilon^*} \cdot \left[\mathrm{LW}(A_k \cdot e) - 2 + \frac{1}{\mathrm{LW}(A_k \cdot e)} \right] \approx \frac{k_B T_0}{\epsilon^*} \cdot [\mathrm{LW}(A_k) - 1], \tag{2}$$

where e = 2.71828, and $\mathrm{LW}(x)$ is the Lambert W function, which asymptotic form is well described by three terms,

$$\mathrm{LW}(z) = \ln(z) - \ln\ln(z) + \frac{\ln\ln(z)}{\ln(z)} + \ldots, \qquad z \gg 1. \tag{3}$$

The SQ efficiency (Eq. 2) is the efficiency for conversion of the radiation power. The thermodynamic efficiency of energy conversion at zero output power is reached in the open circuit regime (negligible current). In the quasi-monochromatic limit, the cell is eliminated by photons within a narrow bandwidth, $\Delta\nu < k_B T_0/h$, around the central frequency ν. For thermodynamic analysis, it is convenient to describe the power of the monochromatic radiation by the temperature T_m. Assuming that the semiconductor bandgap matches the photon energy, we obtained the thermodynamic conversion efficiency at zero output power,

$$\eta_{th} \equiv \frac{qV_{OC}}{\epsilon^*} = \left(1 - \frac{T_0}{T_m}\right) - \ln\frac{1}{k_{ERE}}. \tag{4}$$

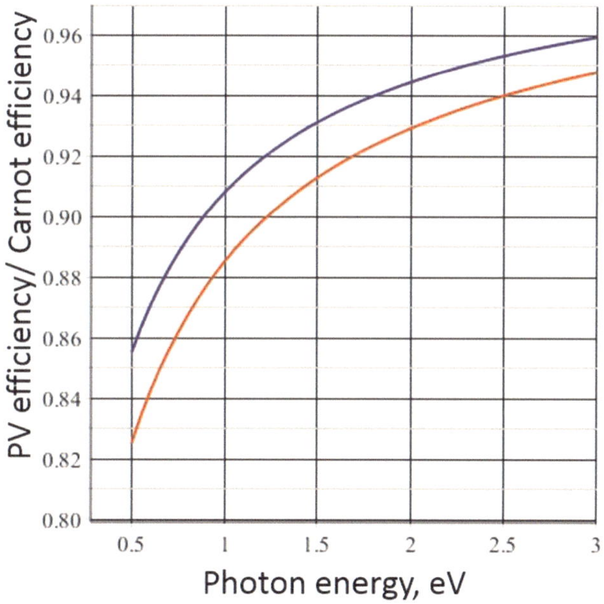

Figure 2.
The limiting detailed-balance SQ monochromatic efficiency normalized by the Carnot efficiency as a function of the photon energy for the monochromatic radiation with the temperature of 1000 K (red) and 10,000 K (blue).

As expected, for the ideal cell ($k_{ERE} = 1$) the thermodynamic efficiency is the Carnot efficiency, $1 - T_0/T_m$ [18].

Analytical solution for the monochromatic energy conversion may be found, when the photons in the incoming flux may be described by the Boltzmann statistics, i.e. $h\nu > k_B T_m$ (for example, for photons with energy of 1.4 eV it means that the photon temperature is limited by 18,000 K). Using Eqs. 1–3 we find the monochromatic SQ conversion efficiency for the ideal cell,

$$\eta_{SQ} = \left(1 - \frac{T_0}{T_m}\right) \cdot \left(1 - \frac{\ln(B)}{B} + \frac{\ln(B)}{B^2}\right)$$
$$B = \frac{h\nu}{k_B T_0} \cdot \left(1 - \frac{T_0}{T_m}\right). \quad (5)$$

This analytical solution is illustrated in **Figure 2**. According to Eq. 5, the detailed-balance SQ conversion efficiency of the monochromatic radiation increases with an increase of the photon energy and approaches the Carnot efficiency at high frequencies.

3. Nonreciprocal monochromatic conversion limit

In this section, we consider the thermodynamic limit and material-determined limit of PV converters with negligible emission realized via nonreciprocal dissipationless photonic management. Let us start with the endoreversible thermodynamics of an engine that receives power from an emitter with temperature T_{em} and operates between temperatures T_{hot} and T_{cold} with the Carnot efficiency, $\eta_{Carnot} = 1 - T_{cold}/T_{hot}$ (**Figure 3a**). It is well understood that for this converter the conversion efficiency of the heat power from the emitter into useful mechanical or electrical power is below the Carnot efficiency due to the emission from the hot sink to the emitter. These emission losses are not directly related to the engine operation. The emission is associated with Kirchhoff's absorption-emission symmetry, which leads to losses in the delivery of heat power from the emitter to the engine. Let us highlight

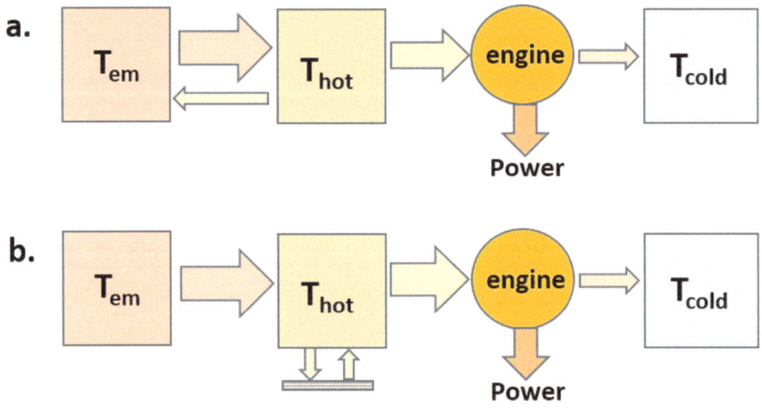

Figure 3.
Conversion of the heat power by the engine that receives heat power from an emitter with temperature T_{em} and operates between heat sinks with temperatures T_{hot} and T_{cold} : (a) a traditional converter with the losses due to emission from the hot sink to the emitter and (b) the nonreciprocal converter with suppressed emission [19, 20].

that Kirchhoff's law is not a thermodynamic law. According to the Onsager - Casimir reciprocity relations, the time-reversal asymmetry in the system may be generated by a magnetic field, magnetization, electric currents, and time-modulation of optical properties [5, 6]. Nonreciprocal optical components can provide absorption channels with near-zero emissivity and emission channels with near-zero absorptivity. As it is shown in **Figure 3b**, we can split absorption channels into two parts. The first part is used for the absorption of incoming light. The second part of the absorption channels is connected with the emission channels and provides 100% reuse of the emitted photons [19]. Also, the emission can be completely suppressed due to quantum effects [20]. Both effects, – the suppressed emission [20] or recycling the emitted radiation to the engine [19], – eliminate the emission losses and increase the conversion efficiency up to the thermodynamic limit given by the Carnot formula.

The above consideration in the frame of endoreversible thermodynamics assumes that the hot end of an engine may be described by the light-increased temperature. Therefore, it does not apply to the nonequilibrium states of semiconductors, which are described by the light-induced chemical potential. Let us also note, that the detailed-balance approach also cannot be employed for the nonreciprocal conversion with zero emission from the cell. Formally, Eq. 2 in this limit gives a divergent result. To determine the nonreciprocal conversion limit we will directly employ the second thermodynamic law. In the general form applicable to non-temperature distributions, the distribution function of photons emitted by electrons in the cell cannot exceed the distribution function of incoming photons. In the quasi-monochromatic limit, we are interested in values of these functions in the narrow bandwidth near the energy hv and, therefore, the limiting value of the light-induced chemical potential may be found from the following equation,

$$\left(\exp\left(\frac{hv}{k_BT_m}\right)-1\right)^{-1}=\left(\exp\left(\frac{hv-\mu}{k_BT_0}\right)-1\right)^{-1}, \tag{6}$$

where T_m is the temperature of the quasi-monochromatic radiation and T_0 is the cell operating temperature. Thus, the thermodynamic limit of the light-induced chemical potential in the nonreciprocal quasi-monochromatic converter is

$$\mu = hv\cdot\left(1-\frac{T_0}{T_m}\right). \tag{7}$$

Taking into account that emission from the nonreciprocal converter is absent and every absorbed photon generates an electron in the output circuit, we see that an ideal nonreciprocal converter provides entire conversion of the photon exergy, $h\nu\cdot(1-T_0/T_m)$. Thus, the conversion efficiency of the nonreciprocal monochromatic converter is the Carnot efficiency,

$$v_{nr} = \left(1-\frac{T_0}{T_m}\right). \tag{8}$$

As the nonreciprocal photonic management provides 100% reuse of the emitted photons, it is interesting to compare the nonreciprocal limiting efficiency (Eq. 8) with the efficiency of the thermophotovoltaic conversion, where the emitted photons are reabsorbed by the emitter and the corresponding energy is reused in conversion. For the monochromatic emitter, the output electric power may be presented as [21],

$$P = \frac{2Hv^3}{c^2} \cdot [N(T_m, \mu = 0) - N(T_0, \mu)] \cdot h\Delta v \cdot \left(1 - \frac{T_0}{T_m}\right), \tag{9}$$

where H is the etendue of incoming and emitted photon bundle (we assume that absorption and emission angles are the same), $N(T_m, \mu = 0)$ is the Bose distribution function of incoming photons with the emitter temperature and zero chemical potential, and $N(T_0, \mu)$ is the Bose distribution function of emitted photons with the cell temperature and light-induced chemical potential. Eq. 9 is the direct consequence of the endoreversible thermodynamics, according to which the limiting output power is given by a product of the Carnot efficiency of the conversion engine and the photon power flux delivered to the engine. The delivered flux is the difference between the absorption and emission fluxes. Assuming that the emitted flux returns to the emitter for later use, the PV conversion efficiency is given by the thermodynamic Carnot efficiency (see Eq. 4.27 in [21]). Thus, the photon reuse via the nonreciprocal photonic management (**Figure 3**) provides the same maximal Carnot efficiency as the photon reuse in the thermophotovoltaic system [21]. Let us highlight, that the Carnot efficiency may be also reached in the PV cell, where the radiative emission processes in the cell are suppressed by quantum effects [20].

Finally, we discuss the material limit of the nonreciprocal (time asymmetric) PV converter. If all emitted photons are reabsorbed by the cell, the radiative recombination lifetime of photocarriers approaches infinity, and photocarrier recombination is realized solely via nonradiative processes. In this case, we can employ the detailed balance SQ approach and corresponding analytical solution given by Eqs. 1 and 2, where the parameter A_k should be changed by

$$A_{nr} = \frac{N_{ab}}{R_{nr}} = A_k \frac{1}{1 - k_{ERE}} \tag{10}$$

where R_{nr} is the equilibrium nonradiative recombination rate, $R_{nr} = n_0 d/\tau_{nr}$, where n_0 is the equilibrium concentration of carriers, d is the cell thickness, and τ_{nr} is the nonradiative recombination time. In the cell design with the nonreciprocal photonic management, effective photon trapping and high photon absorption may be realized via the same external nonreciprocal recycling, which returns radiation to the cell (**Figure 3**). Therefore, the cell thickness may be significantly reduced, which leads to reduction of nonradiative recombination. Also, in high-quality optoelectronic materials the nonradiative recombination time significantly exceeds the radiative time, e.g. in GaAs the ratio $\tau_{nr}/\tau_r \sim 300$ [11]. Thus, the nonreciprocal management with strong suppression of the emission from the GaAs converter can increase the detailed balance coefficient A_k by three orders in magnitude or more, which increases the open circuit voltage at least by 180 mV.

Let us note that emission suppression due to nonreciprocal photonic management strongly enhances photon recycling in the system. The limiting Carnot efficiency of the nonreciprocal converter (Eq. 8) corresponds to the zero-emission and infinite intrinsic photon recycling. Any negligible losses (photon leakage or nonradiative recombination) regularize the solution of the SQ model (Eq. 2). This is a general resolution of thermodynamic paradox related to nonreciprocal power converters and nonreciprocal transfer of electromagnetic energy. In this way, the optical diode paradox was resolved by Ishimaru for the nonreciprocal ferrite-loaded waveguide in 1962 [22]. It was shown that any negligible material loss in ferrites leads to the convergent solution, which does not contradict the second law of thermodynamics [23] (see also a review [5]).

The Ishimarus results and our analysis show that such paradoxes appear, when the Maxwell equations or photon balance equations are applied to a completely lossless medium and systems. Any negligible dissipation eliminates divergent solutions.

4. Photovoltaic greenhouse effect

The Kirchhoff's law is valid for opaque bodies in thermodynamic equilibrium with the environment. As it is highlighted in Ref. [24], these two assumptions are often not satisfied. In particular, any photovoltaic converter operates in strongly nonlinear regime far from equilibrium. In this section, we investigate the nonreciprocal photonic management realized due to greenhouse type filter, which mimics the greenhouse operation, where the greenhouse glass/plastic reduces thermal emission and preserves more thermal energy in the greenhouse. The greenhouse filter is placed at the front surface of the cell. It reflects low energy photons in some narrower energy interval ($\sim$ 50–150 meV) above the semiconductor bandgap (see **Figure 1**). In other words, the filter and back surface mirror establishes the photonic bandgap, σ_{ph}, above the semiconductor bandgap, σ_g. The greenhouse filter prevents escape of photons with energies in the range between semiconductor and photonics bandgaps. The filter separates main electronic and photonic processes in the PV converter. Absorption-emission balance and conversion efficiency of this converter drastically depend on the light-induced concentration of hot photocarriers above the photonic bandgap established by the greenhouse filter. For the monochromatic conversion, the photonic bandgap, should correspond to the energy of radiation quanta, $\sigma_{ph} = hv$. As the photonic bandgap exceeds the semiconductor bandgap by $\sim$100–150 meV, the absorption at the photonic bandgap strongly exceeds the absorption near the semiconductor bandgap. To achieve negligible nonreciprocal emission above the photonic bandgap, the population of hot electrons with energy above the photonic bandgap should be strongly reduced by fast cooling processes that dominate over thermo-excitation processes of photocarriers accumulated above the semiconductor bandgap, σ_g. In the limit of strong greenhouse effect, i.e. negligible emission above the photonic bandgap, we can repeat consideration of the previous section and obtain the limiting conversion efficiency of the greenhouse PV converter,

$$\eta_{gh} = \frac{\sigma_g}{hv}\left(1 - \frac{T_0}{T_m}\right) = \frac{\sigma_g}{\sigma_{ph}}\left(1 - \frac{T_0}{T_m}\right). \tag{11}$$

The limiting efficiency of the greenhouse PV converter is smaller than the Carnot efficiency by the factor of σ_g/σ_{ph}, because the nonreciprocity is realized via nonreversible dissipative processes. As all kinetic processes are essentially reversible in equilibrium, the nonreversible relaxation of photocarriers between photonic bandgap and semiconductor bandgap levels requires that the level separation substantially exceeds the thermal energy, i.e. $\sigma_{ph} - \sigma_g \geq 2 - 3k_BT_0$. In other words, the dissipated energy should exceed $\sim 2 - 3k_BT_0$ per photocarrier to reach electromagnetic nonreciprocity via relaxation processes. Without such separation, we will return to the results of Ref. [9], which show that the PV conversion is insensitive to photocarrier kinetics within the thermal energy scale.

As we discussed above, the efficiency of the greenhouse PV converter strongly depends on cooling mechanisms of hot photocarriers. Let us consider kinetics of the

GaAs cell. The photon with energy above the semiconductor band gap creates electron and hole with energies

$$E_e = (hv - \sigma_g) \cdot \frac{m_h}{m_h + m_e} \tag{12}$$

$$E_h = (hv - \sigma_g) \cdot \frac{m_e}{m_h + m_e}. \tag{13}$$

Usually in semiconductor materials, the whole mass strongly exceeds the electron mass and practically whole photon energy is transferred to photoelectron. In GaAs these effective masses are $m_e = 0.067m_0$ and $m_h = 0.45m_0$, where m_0 is the free electron mass. Therefore, to manage kinetics of hot photocarriers we should choose the p-doped GaAs.

Photoelectrons accumulated above semiconductor bandgap may be described by the Boltzmann distribution function with the light-induced chemical potential μ_{sc},

$$f(\epsilon \approx \sigma_g) = \exp\left(\frac{\mu_g - \epsilon}{k_B T_0}\right) \tag{14}$$

These photoelectrons are collected and produce the output voltage $V = \mu_g/q$. In the narrow range of the order of thermal energy above the photonic bandgap, σ_{ph}, the distribution function may be approximated by the chemical potential μ_{ph},

$$f(\epsilon \approx \sigma_{ph}) = \exp\left(\frac{\mu_{ph} - \epsilon}{k_B T_0}\right) \tag{15}$$

If inter-electron interaction dominates over other processes in photoelectron kinetics, the whole system is described by the same chemical potential. The conversion efficiency is given by Eq. 2 with the absorption-emission balance established above the photonic bandgap. In this case, the greenhouse filter only suppresses the matching losses. If cooling of photoelectrons above the photonic bandgap dominates over phonon-induced thermo-excitation of photoelectrons (see **Figure 4**), the

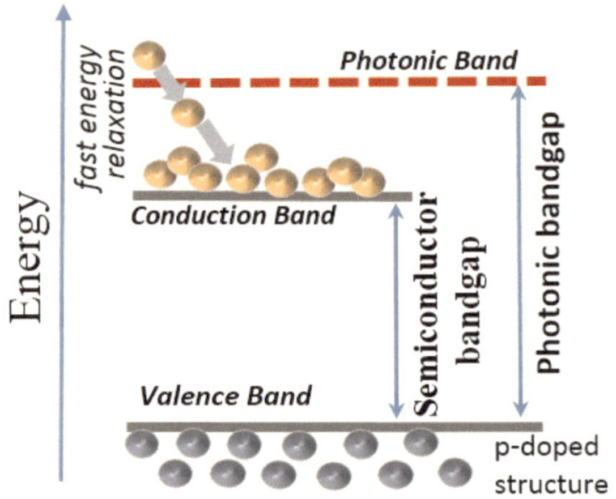

Figure 4.
Fast energy relaxation of the beam-generated photoelectrons depopulates energy levels above the photonic bandgap. As a result, the chemical potential of hot electrons and emitted photons that leave converter is much less than the chemical potential of photoelectrons accumulated near the semiconductor bandgap.

chemical potential of hot photoelectrons, μ_{ph}, is much smaller than μ_g. A reduced population of hot photoelectrons directly reduces the generation of photons with energy above the photonic bandgap. Because only these photons can leave the converter, emission from the converter is suppressed by the factor of $\exp\left(\mu_g/\mu_{ph}\right)$..

Typically, cooling processes of hot photoelectrons are rather fast and, therefore, the chemical potential μ_{ph} is much smaller than μ_g. Cooling processes are exceptionally strong in A_3B_5 semiconductors [16, 25–28], where the valence band consists of light and heavy holes, and photo-induced hot electrons transfer their energy to heavy holes created by p-doping. The electron cooling may be realized via inelastic scattering of electrons on heavy holes that is accompanied by the transition of a heavy hole to the light-hole band [25] as well as via inelastic electron scattering on oscillations of multicomponent hole plasma with heavy and light holes [28]. Evaluations show that the many body processes dominate in hot carrier cooling. Using the results of Ref. 28, in our modeling, we employ the following equation for the energy relaxation rate of hot electrons due to energy transfer to hole plasmons,

$$\frac{1}{\tau_{e-h}} = \frac{\pi \hbar^2 \cdot p_0}{m_{lh}^{3/2} \sqrt{\epsilon_e}} \tag{16}$$

where p_0 is the hole concentration due to doping, m_{lh} is the mass of the light hole, and ϵ_e is the electron energy. Eq. 16 applies to electrons with $\epsilon_e \geq 15$ meV above the semiconductor bandgap for the doping levels below 310^{18} cm^{-3}. The energy transfer from hot electrons to holes is not limited by thermal energy. The characteristic transferred energy changes from $5\hbar^2 p_0^{3/2}/m_{hh}$ (m_{hh} is the mass of the heavy hole) to $5\hbar^2 p_0^{3/2}/m_{lh}$ and substantially exceeds thermal energy. As a result, the electron energy relaxation due to interaction with hole plasma oscillations is very fast. Femtosecond electron cooling in GaAs and other A_3B_5 materials was observed in numerous experimental investigations [16, 29–31]. The femtosecond electron–hole relaxation strongly dominates over electron–phonon processes with a characteristic time of 1–2 ps.

To illustrate the operation of the greenhouse converter we perform simulations of conversion efficiency of 10W/cm^2 laser radiation with a wavelength around 809 nm by GaAs cell with the greenhouse filter that establishes the photonic bandgap equaled to the laser energy quanta. The integration of GaAs cell with this laser is widely studied for power beaming [32]. To enhance absorption we add the Lambertian scattering layer placed between the filter and the cell (the same effect may be reached by the curved or textured mirror). Lambertian scattering significantly enhances light absorption by the cell. The corresponding absorption coefficient is given by.

$$A = \frac{1 - \exp(-4\alpha d)}{1 - (1-b)(1-1/n^2) \cdot \exp(-4\alpha d)}, \tag{17}$$

where d is the cell thickness, α is the GaAs absorption at the laser wavelength, n is the refractive index, and b is the absorption of the back surface mirror.

The modeling of PV performance is based on analytical SQ solution (Eqs. 1 and 2), which was described in details for GaAs cell in Ref. cite4 and for various thermophotovoltaic cells in Ref. [33]. In Ref. [3] we investigated the same power beaming conversion in quasi-equilibrium approximation $\mu_{ph} = \mu_g$. In this modeling, we take into account the greenhouse PV effect due to fast electron cooling and, using the Boltzmann equation we calculate the reduced emission flux.

The main results of our modeling are presented in **Figure 5**, which shows the increase in the conversion efficiency due to the greenhouse filter as a function of internal quantum efficiency (IQE), k_{int} and normalized cell thickness, $d\alpha$. As seen, the increase in efficiency can reach 12–14%, but significant improvements require high-quality PV materials with high internal quantum efficiency.

To distinguish the nonequilibrium greenhouse effect in PV conversion, in **Figure 6** we present the conversion efficiency calculated in the nonequilibrium model with fast photoelectron cooling, quasi-equilibrium approximation, and for traditional cell design without the greenhouse filter. The efficiency as a function of the dimensionless cell thickness, $d\alpha$, is shown for several values of the internal quantum efficiency, k_{int}. For available GaAs materials, the IQE may reach 0.997 [11] and the greenhouse filter adds 7% to the conversion efficiency, where 1% is added due to nonequilibrium effects (red, blue, and green solid lines). At $k_{int} = 0.998$ (yellow line) the efficiency exceeds the SQ power beaming efficiency (blue dashed line). The limiting efficiency (green dashed line) exceeds the SQ efficiency by 7%.

Figure 7 demonstrates the conversion efficiency of 809 nm laser radiation as a function of the internal quantum efficiency (**Figure 7a**) and the laser power (**Figure 7b**) for the greenhouse GaAs PV converter (green lines), the quasi-equilibrium approximation for the same converter (blue lines), and traditional converter without the greenhouse filter. As it is shown in **Figure 7b**, all converters have the same, weak (logarithmic) dependence of the efficiency on the laser power. Dependencies of efficiency on the material IQE are substantially different. While the efficiency of the traditional converter has a rather weak, linear dependence on IQE, the performance of the greenhouse converter in the quasi-equilibrium approximation and especially in the model with photoelectron cooling strongly depends on the IQE.

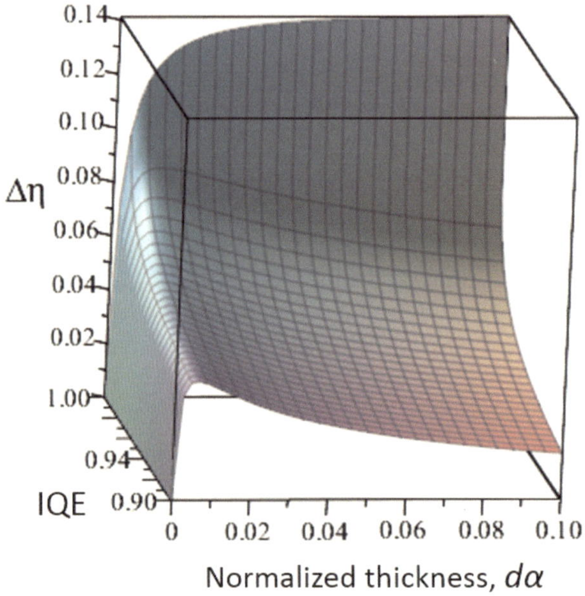

Figure 5.
Increase in the conversion efficiency of the laser light with 809 nm wavelength and power of 10 W/cm² due to the PV greenhouse effect as a function of internal quantum efficiency (IQE) and normalized cell thickness, dα.

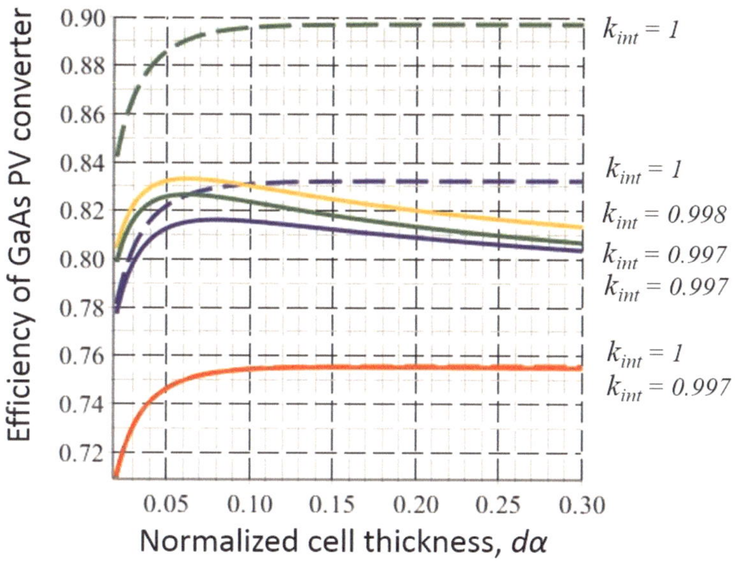

Figure 6.
PV efficiency of GaAs device, which converts laser light with 809 nm wavelength and power of 10 W/cm² vs. cell thickness. Cell without greenhouse filter: Red lines (dashed $k_{int} = 1$, solid $k_{int} = 0.997$); cell with the filter in quasi-equilibrium approximation: Blue lines (dashed $k_{int} = 1$, solid $k_{int} = 0.997$); cell with the filter and the greenhouse kinetics: Green lines (dashed $k_{int} = 1$, solid $k_{int} = 0.997$) and yellow line ($k_{int} = 0.998$)..

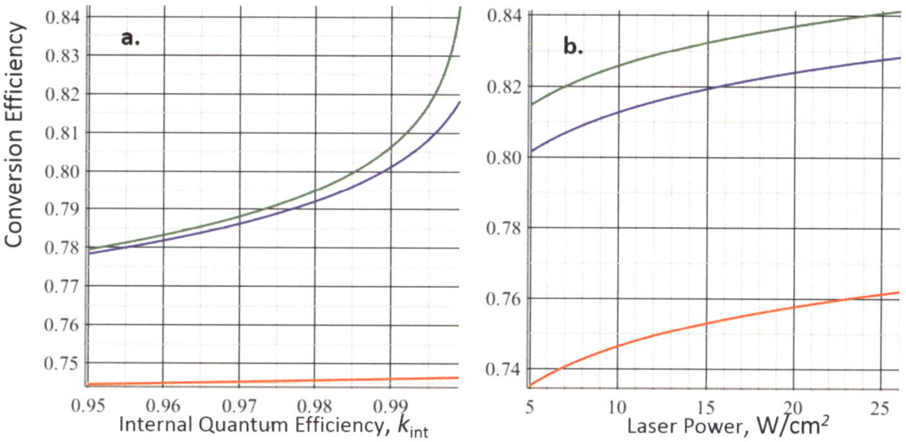

Figure 7.
Increase in the conversion efficiency of the laser light with 809 nm wavelength and power of 10 W/cm² due to the greenhouse PV effect as a function of internal quantum efficiency (IQE) and normalized cell thickness, dα.

Our modeling shows that for suppression of the laser-cell matching losses (the quasi-equilibrium model) we need good PV materials with IQE better than 0.8. The efficiency improvement due to suppression of emission losses via cooling of photoelectrons requires high-quality PV materials with IQE better than 0.99. Otherwise, the nonradiative recombination dominates over the radiative component and determines the converter performance.

Perovskites are low-cost and rather nonhomogeneous materials. These materials show a very smooth absorption edge with the width ~ 100 nm [14]. While above this range perovskites demonstrate excellent, very close to unity EQE, the smooth absorption drastically reduces the photovoltaic performance [14, 15]. Rau et al. [34] proposed that perovskites may be considered as a semiconductor with the distributed PV bandgap [35]. This model is widely applied to perovskite cells and successfully explains a significant reduction of conversion efficiency with respect to SQ limit [34–36]. In particular, according to the distributed bandgap model, the 100 meV standard deviation from the mean bandgap reduces the conversion efficiency by 6%. The greenhouse filter is a valuable tool to establish the sharp absorption edge above the smooth material absorption edge. Thus, for the power beaming with perovskite cells, the greenhouse design is expected to increase the conversion efficiency due to the suppression of both the laser-cell matching losses and the distributed bandgap losses.

5. Conclusions

Photonic management of radiative processes is an effective tool to enhance the performance of photovoltaic converters. An ideal nonreciprocal converter provides the entire conversion of the photon exergy, $h\nu \cdot (1 - T_0/T_m)$. Nonreciprocal management is the most radical way to change the absorptionemission balance in favor of the absorption. Suppression of spontaneous emission via quantum interference effects was proposed in seminal works of Scully to increase the PV efficiency of quantum photocell [20]. Here we have proposed and investigated a more practical way to suppress emission due to a narrow bandwidth filter, which generates the greenhouse effect in a cell. As an ordinary greenhouse effect, the greenhouse photovoltaic effect requires fast cooling of photocarriers and strong trapping of low-energy photons, which are emitted by photoelectrons near the semiconductor bandgap. In A_3B_5 semiconductors, effective electron cooling is realized by energy transfer from hot photoelectrons to the plasma oscillations of holes. Let us note, that the greenhouse effect may be also realized in PV design with a 3D photonic crystal, which has a photonic bandgap that overlaps the electronic band edge. In this design, the emission inside the photonic bandgap is rigorously forbidden [37]. Greenhouse photonic management has a strong potential to substantially increase the conversion efficiency due to the reduction of the laser-cell matching losses, the radiative losses, and the distributed bandgap losses in low-cost perovskites and organic materials. Nonreciprocal photovoltaics requires materials with high internal quantum efficiency and high-quality optical components, which provide enhanced photon recycling.

Acknowledgements

The work is supported by the Army Research Laboratory. Research of AS was accomplished under Cooperative Agreement No. W911NF-18-2-0222. The work of KS is supported by Texas A&M University. The views and conclusions contained in this paper are those of the authors and should not be interpreted as representing the official policies, either expressed or implied, of the Army Research Laboratory or the US Government.

Author details

Andrei Sergeev[1*†] and Kimberly Sablon[2,3†]

1 U.S. Army Research Laboratory, Adelphi, Maryland, USA

2 Office of Undersecretary of Defense for Research and Engineering, Washington, DC, USA

3 Bush Combat Development Complex, Texas A&M University, Bryan, Texas, USA

*Address all correspondence to: podolsk37@gmail.com

† These authors contributed equally.

References

[1] Hohn O, Walker AW, Bett AW, Helmers H. Optimal laser wavelength for efficient laser power converter operation over temperature. Applied Physics Letters. 2016;**108**:241104

[2] Algora C, Garcia I, Delgado M, et al. Beaming power: Photovoltaic laser power converters for power-by-light. Joule. 2022;**6**:340-368

[3] Sergeev A, Hewitt A, Hier H, et al. Greenhouse effect in photovoltaic cells to enhance efficiency of power beam conversion. MRS Advances. 2019;**4**: 897-903

[4] Sablon KA, Sergeyev AV, and Little JW. High Efficiency Photovoltaic Cells with Suppressed Radiative Emission Due to Chemical Nonequilibrium of Photoelectrons. US Patent # 11,152,888 (2021)

[5] Caloz C, Alu A, Tretyakov S, et al. Electromagnetic nonreciprocity. Physical Review Applied. 2018;**10**: 047001

[6] Jalas D, Petrov A, Eich M, et al. What isand what is notan optical isolator. Nature Photon. 2013;**7**:579582

[7] Fourier J. Thorie Analytique de la Chaleur (the Analytical Theory of Heat), 1822

[8] Shockley W, Queisser HJ. Detailed balance limit of efficiency of p-n junction solar cells. Journal of Applied Physics. 1961;**32**:510

[9] Kamide K, Mochizuki T, Akiyama H, Takato H. Nonequilibrium theory of the conversion efficiency limit of solar cells including thermalization and extraction of carriers. Physical Review Applied. 2018;**10**:044069

[10] Green MA, Bremner SP. Energy conversion approaches and materials for high-efficiency photovoltaics. Nature Materials. 2017;**16**:2334

[11] Schnitzer I, Yablonovitch E, Caneau C, Gmitter TJ. 30% external quantum efficiency from surface textured, thinfilm lightemitting diodes. Applied Physics Letters. 1993;**62**: 131

[12] Green MA, Ho-Baillie AWY. Pushing to the limit: Radiative efficiencies of recent mainstream and emerging solar cells. ACS Energy Letters. 2019;**4**: 16391644

[13] Green MA. Radiative efficiency of state-of-the-art photovoltaic cells. Progress in Photovoltaics: Research and Applications. 2012;**20**:472476

[14] Tress W. Perovskite solar cells on the way to their radiative efficiency limitinsights into a success story of high opencircuit voltage and low recombination. Advanced Energy Materials. 2017;7:1602358

[15] Brinkmann KO, Becker T, Zimmermann F, et al. The optical origin of near-Unity external quantum efficiencies in perovskite solar cells. Solar RRL. 2021;**5**:2100371

[16] Hohenester SUP, Kocevar P, et al. Subpicosecond thermalization and relaxation of highly photoexcited electrons and holes in intrinsic and p-type GaAs and InP. Physical Review B. 1993;**47**:13233-13245

[17] Sergeev A, Sablon K. Exact solution, endoreversible thermodynamics, and kinetics of the generalized Shockley-Queisser model. Physical Review Applied. 2018;**10**:064001

[18] Rau U, Paetzold UW, Kirchartz T. Thermodynamics of light management in photovoltaic devices. Physical Review B. 2014;**90**:035211

[19] Sergeev A, Sablon K. Nonreciprocal photonic management for photovoltaic conversion: Design and fundamental efficiency limits. Journal of Photonics for Energy. 2022;**12**:032207

[20] Scully MO. Quantum photocell: Using quantum coherence to reduce radiative recombination and increase efficiency. Physical Review Letters. 2010;**104**:207701

[21] Luque A, Marti A. Theoretical limits of photovoltaic conversion. In: Luque A, Hegedus S, editors. Handbook of Photovoltaic Science and Engineering. Chichester, UK: Wiley; 2003

[22] Ishimaru A. Unidirectional Waves in Anisotropic Media and the Resolution of the Thermodynamic Paradox. Seattle: Tech. Rep., Washington University; 1962

[23] Ishimaru A. Electromagnetic Wave Propagation, Radiation, and Scattering: From Fundamentals to Applications. 2nd ed. Wiley IEEE Press; 2017

[24] Tanyi EK, Burton BT, Narimanov EE, Noginov MA. Thermal radiation of Er doped dielectric crystals: Probing the range of applicability of the Kirchhoff's law. Scientific Reports. 2017;7:2040

[25] Dyakonov MI, Perel VI, Yassievich IN. Effective mechanism of energy relaxation of hot-electrons in p-type semiconductors. Soviet Physics – Semiconductors. 1997;**11**:801-804

[26] Joshi RP, Grondin RO, Ferry DK. Monte Carlo simulation of electron-hole thermalization in photoexcited bulk semiconductors. Physical Review B. 1990;**42**:5685

[27] Rota L, Ferry DK. Monte Carlo investigation of carrier-carrier effects in femtosecond pump and probe experiments. Semiconductor Science and Technology. 1994;**9**:468-470

[28] Fedirko VA, Zakharova AA. Hot electron relaxation due to equilibrium density fluctuations in a hole plasma. Soviet Physics – Semiconductors. 1991; **25**:970-973

[29] Young JF, Kelly PJ, Henry NL. Quantitative evaluation of the energy-loss rate of hot electrons to cool electron-hole plasmas including dynamic screening and intervalence band processes. Semiconductor Science and Technology. 1992;7:8148-8150

[30] Gong T, Fauchet PM, Young JF, Kelly PJ. Subpicosecond hot hole dynamics in highly excited GaAs. Applied Physics Letters. 1993;**62**:522-524

[31] Rodrigues-Herzog R, Sailer M, Hecker NE, et al. Ultrafast energy loss of electrons in p-GaAs. Applied Physical Letters. 1995;**67**:264

[32] Oliva E, Dimroth F, Bett AW. GaAs converters for high power densities of laser illumination. Progress Photovoltaics: Research and Applications. 2008;**16**:289

[33] Sergeev A, Waits CM. Effects of photon recycling, trapping, and reuse on thermophotovoltaic conversion efficiency and output power. Journal of Photonics for Energy. 2020;**10**:035501

[34] Rau U, Blank B, Muller TCM, Kirchartz T. Efficiency potential of photovoltaic materials and devices unveiled by detailed-balance analysis. Physical Review Applied. 2017;7:044016

[35] Rau U, Werner JH. Radiative efficiency limits of solar cells with lateral

band-gap fluctuations. Applied Physics Letters. 2004;**84**:3735-3737

[36] Nayak PK, Mahesh S, Snaith HJ, Cahen D. Photovoltaic solar cell technologies: Analysing the state of the art. Nature Reviews Materials. 2019;**4**: 269285

[37] Yablonovitch E. Inhibited spontaneous emission in solid-state physics and electronics. Physical Review Letters. 1987;**58**:2059-2062

Chapter 4

Comparison of Evaporation in Conventional Diesel and Bio-Fuel Droplets in Engine Cylinder

Ali Raza, Zunaira-Tu-Zehra, Muhammad Khurram, Muhammad Ahsan Pervaiz Khan, Asif Durez and Liaquat Ali Khan

Abstract

Renewable energy resources are need of the hour at the current energy scarcity scenario in the world. Scientist and researchers are finding the ways to replace the conventional energy resources with the renewable ones. It is fact that fossils are going to be obsolete in future. One third of global energy is being consumed by the transportation sector. All the amount of this energy comes from the fossils that contain the hydrocarbons in their composition. Efforts are being made to replace the fossils with the renewable energy resources. In this regard, biofuels are emerged as a replacement of the diesel fuels. There are several processes in the engine cylinder from atomization of fuel until the exhaust of gases. One of them is the evaporation of fuel droplets. In the present work, evaporation characteristics of conventional diesel fuel and biofuels is described by comparing them in different working conditions. Modeling of evaporation phenomenon using computational fluid dynamics (CFD) techniques and the effects of in cylinder conditions is also explained. Results show that biofuel droplets show a better evaporation rate at the high operating conditions in the engine cylinder.

Keywords: diesel, biofuel, droplet, computational fluid dynamics, evaporation, renewable energy

1. Introduction

Renewable energy refers to energy sources that are replenished naturally and continuously in a relatively short period, unlike non-renewable sources such as fossil fuels. Renewable energy sources include solar, wind, hydro, geothermal, and biomass, among others. These sources of energy are considered renewable because they can be replenished over time through natural processes, such as the sun's energy replenishing solar panels or wind turbines, or the regrowth of crops for use in biomass energy. This makes renewable energy a sustainable alternative to traditional non-renewable energy sources, which will eventually run out and can have negative impacts on the environment [1]. Renewable energy sources are sources of energy that are replenished naturally and continuously in a relatively short period of time [2]. Some of the most common sources of renewable energy include:

1. Solar energy: Energy derived from the sun by solar panels or other devices that convert sunlight into electrical or thermal energy [3].

2. Wind energy: Energy generated by the movement of wind, typically using wind turbines [4].

3. Hydro energy: Energy generated by the movement of water, typically with hydroelectric dams or other water-based power generation systems [5].

4. Geothermal energy: Energy derived from the heat produced by the Earth's core, typically through the use of geothermal power plants [6].

5. Biomass energy: Energy generated from organic matter, such as crops, wood, and waste products, through the process of combustion, anaerobic digestion, or other means [7].

6. Tidal energy: Energy generated from the movement of tides, typically by tidal turbines or other ocean energy systems [8].

7. Biofuels, which are fuels made from biological materials such as crops and waste [9].

All these sources of renewable energy are sustainable and do not deplete the earth's resources or produce harmful greenhouse gas emissions.

There are several reasons why we need renewable energy:

1. Sustainable energy supply: Renewable energy sources are replenished naturally and continuously, making them a sustainable alternative to non-renewable sources of energy, such as fossil fuels, which will eventually run out [10].

2. Reduction of greenhouse gas emissions: The use of renewable energy sources, such as solar and wind power, results in fewer greenhouse gas emissions, which contribute to climate change, compared to the use of non-renewable energy sources, such as coal and natural gas.

3. Improved energy security: Renewable energy sources are typically dispersed and decentralized, making them less vulnerable to disruption and more resilient in the face of natural disasters or political conflicts [11].

4. Job creation: The development and deployment of renewable energy technologies can create new jobs in the fields of engineering, manufacturing, construction, and installation, among others.

5. Economic benefits: Renewable energy sources can reduce the cost of energy over time, as the cost of renewable energy technologies continues to decline. They can

also stimulate local economies by attracting investment and reducing the need for expensive energy imports [12].

6. Health benefits: The use of renewable energy sources can reduce air and water pollution, leading to improved public health and reduced healthcare costs [13].

Overall, renewable energy can help address many pressing global challenges such as climate change, energy security, job creation, and sustainable development. Fossils are the preserved remains or evidence of ancient plants, animals, or other organisms that lived in the distant past. They can take many forms, including bones, teeth, shells, footprints, and even impressions of leaves or insects. Fossils are typically found in sedimentary rock, such as sandstone or limestone, but can also be found in volcanic ash or ice. Fossilization is the process by which the remains of ancient organisms become fossils. This process can occur through a variety of mechanisms, including preservation in sediment, freezing in ice, or preservation in amber. In most cases, minerals replace the original organic material over time, creating rock-like fossils [14].

Fossils are important for several reasons:

1. They provide evidence of past life on Earth, including information about the diversity, distribution, and evolution of ancient organisms.

2. They can be used to date the rock layers in which they are found, providing important information about the Earth's history and geology.

3. They provide insight into ancient environments, including information about past climate and the types of organisms that lived in a particular area.

4. Fossil fuels such as natural gas, coal, and oil, on the other hand, are formed from the remains of animals and plants that were buried under sediment and were then subjected to heat and pressure over millions of years [15].

There are several types of fossils, including:

1. Body fossils: Body fossils are the remains of an organism's physical structure, such as bones, shells, and teeth. They provide direct evidence of the anatomy and biology of ancient organisms and are often used to study the evolution and diversity of species over time.

2. Trace fossils: These are the indirect evidence of ancient organisms, such as footprints, burrows, or coprolites (fossilized feces).

3. Molds and casts: These are formed when a fossilized organism dissolves away, leaving a negative impression or mold in the rock. A cast is formed when this mold is filled with mineral deposits.

4. Petrified fossils: These are formed when mineral deposits replace the original organic material of a fossil, creating a rock-like fossil.

5. Carbon films: These are extremely thin layers of carbon that have been left behind by the decay of organic material. They can reveal the fine details of an organism's structure.

6. Amber fossils: These are formed when organisms are trapped and preserved in tree sap, which hardens into amber over time.

7. Per mineralized fossils: These are formed when minerals fill the spaces within bones and shells, preserving them in great detail.

8. Original remains: These fossils still have their original materials such as bone, teeth, shells, etc. [16].

All these types of fossils can provide important information about ancient organisms and environments and have a wide range of scientific applications. The lifetime of fossils can vary widely depending on the type of fossil and the conditions under which it formed. Some fossils, such as those found in amber, could be well preserved for millions of years, while others, such as those found in surface sediments, may only be preserved for a few thousand years. The preservation potential of a fossil also depends on the type of organism, the environment in which it lived, and the conditions under which it died. For example, organisms that lived in environments with high sedimentation rates, such as near river deltas or oceanic currents, are more likely to be buried and preserved as fossils than those that lived in environments with low sedimentation rates. Similarly, organisms that are rapidly buried after death, such as those that are transported to the bottom of the ocean by a volcanic eruption or landslide, are more likely to be preserved as fossils than those that are exposed to the elements for long periods of time before being buried [17].

In general, the most common types of fossils are body fossils, such as bones and shells, and trace fossils, such as footprints, which can be preserved for millions of years under the right conditions. Other types of fossils, such as carbon films and original remains, may only be preserved for a few thousand years [18]. Fossil fuels such as natural gas, coal, and oil, on the other hand, are formed from the remains of animals and plants that were buried under sediment and were then subjected to heat and pressure over millions of years. The process of fossilization can take millions of years, and once formed, these fossil fuels can be stored in the Earth's crust for millions of years more [19].

The combustion of fossil fuels, such as coal, oil, and natural gas, can have a range of negative environmental effects. Some of the most significant effects include:

1. Greenhouse gas emissions: The burning of fossil fuels releases carbon dioxide (CO_2) and other greenhouse gases into the atmosphere, which contribute to global warming and climate change.

2. Air pollution: The burning of fossil fuels also releases a range of pollutants into the air, such as sulfur dioxide, nitrogen oxides, and particulate matter, which can harm human health and the environment.

3. Water pollution: Fossil fuel extraction and processing can also pollute water resources, including the contamination of surface and ground water with chemicals and heavy metals.

4. Land degradation: Fossil fuel extraction can cause land degradation, including the destruction of natural habitats, deforestation, and soil erosion [20].

5. Health impacts: Air pollution from fossil fuel combustion can lead to respiratory and cardiovascular disease, as well as cancer [21].

6. Acid rain: The burning of fossil fuels release sulfur dioxide and nitrogen oxides, which can lead to acid rain, which can harm plants, animals, and buildings.

7. Oil spills: The transportation of oil and oil products by ships and pipelines can lead to oil spills, which can have devastating effects on marine life and coastal environments [20].

Overall, the combustion of fossil fuels can have severe environmental impacts and contributes to climate change, air pollution, water pollution, land degradation and health impacts.

Biofuels are fuels that are derived from biomass, or organic matter such as plants, crops, and waste materials. They are renewable energy sources, as they can be replenished on a regular basis [21]. There are several types of biofuels, including:

1. Bioethanol: This fuel is made from fermented crops, such as corn, sugarcane, wheat. It is often used as a gasoline additive which is being used in order to increase fuel efficiency and reduce greenhouse gas emissions.

2. Biodiesel: A diesel fuel type which is actually extracted from vegetable oils, animal fats, or recycled cooking oils. It is also used in conventional diesel engines.

3. Biogas: This is a type of fuel made from the decomposition of organic matter, such as food waste, manure, or sewage. It can be used to generate heat, electricity, or as a transportation fuel.

4. Biojet fuel: This fuel is extracted from plant-based oils or animal fats, specifically designed for use in aircraft.

Biofuels have several benefits, including minimizing the reliance on the fossil fuels, bring down carbon emissions, creating new jobs in agriculture and biofuels production, and improving energy security by reducing the reliance on imported oil. However, there are also some potential drawbacks to biofuels, including the competition for land and water resources with food production, the environmental impact of growing crops specifically for biofuels, and the fact that some biofuels may not be as energy efficient as conventional fossil fuels [22].

The composition of biofuels depends on the specific type of biofuel in question. However, some of the most common components of biofuels include:

1. Ethanol: Ethanol is extracted from crops such as corn, sugarcane, and wheat. It is often used as a gasoline additive which is being used in order to increase fuel efficiency and reduce greenhouse gas emissions.

2. Methanol: Methanol is another type of alcohol that can be made from biomass, such as wood waste, and is used as a fuel or as a starting material for the production of other chemicals.

3. Vegetable oils: Biodiesel is often made from vegetable oils, including soybean oil, palm oil, or canola oil.

4. Animal fats: Biodiesel can also be made from animal fats, such as tallow or chicken fat.

5. Glycerol: Glycerol is a byproduct of the production of biodiesel and is commonly used as a feedstock to produce other chemicals.

6. Biogas: Biogas is actually mixtures of hydrocarbon gases which mainly includes the mixture of methane and carbon dioxide that is produced through digestion of organic matter in absence of oxygen, such as food waste, manure, or sewage.

7. Algal oils: Algal oils are oils that are produced from microalgae and can be used as feedstocks to produce biofuels [23].

The composition of biofuels can vary depending on the specific feedstock used and the production process, but in general, they are composed of organic compounds such as alcohols, acids, and esters. These organic compounds have a lower carbon content and release less carbon dioxide when burned compared to fossil fuels, making them a more sustainable and environmentally friendly alternative [24].

Following are the resources of biofuels:

1. Agricultural crops: Biofuels such as bioethanol and biodiesel which are mainly present in sugarcane, soybeans, and canola.

2. Forest residues: Biofuels can be made from waste products generated by the forest industry, such as sawdust and bark.

3. Food waste: Organic waste is another main source such as food waste, manure, and sewage.

4. Algae: Algal biofuels are produced from microalgae and can be used to produce biofuels such as biodiesel and bio jet fuel.

5. Cellulosic crops: Biofuels can be produced from crops that are high in cellulose, such as switchgrass, corn stalks, and sugarcane bagasse.

6. Animal fats: Biodiesel can be made from animal fats such as tallow and chicken fat.

Overall, the range of biofuels sources is quite diverse, and the specific sources used will depend on a range of factors, including the availability of feedstocks, the cost of production, and the specific type of biofuel being produced [25].

Following are the processes that occur in the cylinder:

a. Cavitation in nozzle

b. Primary break up

c. Secondary break up

d. Evaporation of fuel droplets

e. Combustion of fuel droplets

f. Exhaust effect

Cavitation is a phenomenon that can occur in nozzles, particularly in high-speed fluid flow applications. It occurs when the pressure of the fluid become less than the vapor pressure, which eventually resulted in bubble formation. These bubbles can then collapse, and released energy like in the form of shock waves, turbulence, and high-velocity fluid jets. This can cause damage to the nozzle and reduce the efficiency of the fluid flow. In a nozzle, cavitation can happen when the fluid pressure become lower than its vapor pressure due to the high velocity of the fluid as it flows through the nozzle. This can result in the formation of vapor bubbles, which can then collapse and cause damage to the nozzle and surrounding areas. To prevent cavitation, the design of the nozzle must be optimized to maintain a high fluid pressure, or the fluid must be kept at a temperature above its boiling point. Other methods of preventing cavitation include adding a suction device, such as a venturi, to increase the fluid pressure, or reducing the fluid velocity by reducing the flow rate or increasing the nozzle diameter.

It can cause damage to the nozzle and reduce the efficiency of fluid flow, so it is important to prevent or mitigate cavitation in high-speed fluid flow application [26]. Primary breakup refers to the initial fragmentation of a fluid stream into smaller droplets or particles. This process occurs when a fluid stream is forced through a small opening or nozzle, or when it is subjected to high-velocity flow, turbulence, or other forms of shear stress. Primary breakup is important in various steps during industrial processes, such as spray dying, atomization, and fuel injection, as well as in natural phenomena such as rain formation. The dispersion of fuel droplets produced during primary breakup can create impact on the efficiency and effectiveness of the process. To control and optimize primary breakup, a variety of techniques can be used, including the design of the nozzle or other dispersion device, the addition of surfactants or other agents to reduce surface tension and promote droplet formation, and the control of fluid velocity and turbulence [27]. Secondary breakup refers to the fragmentation of droplets into much smaller droplets that occurs after the initial primary breakup of a fluid stream. This process typically occurs when the primary droplets collide or interact with each other, leading to further fragmentation and the formation of smaller droplets. Secondary breakup is an important factor in many industrial processes, such as spray drying, atomization, and fuel injection, as well as in natural phenomena such as rain formation. To control and optimize secondary breakup, the techniques are same as mentioned in primary break up process, including the modifications in nozzle design, the control of fluid velocity and turbulence, and the use of additives to modify the physical properties of the fluid [28]. Evaporation of fuel droplets refers to the process by which the liquid fuel droplets present in an engine's combustion chamber are converted into vapor. This process is an essential part of the internal combustion engine, as it allows the fuel to mix with air and burn to produce power. Fuel droplets are formed during the atomization process, which occurs when the liquid fuel is forced through a small opening, such as a fuel injector, to produce a spray of fine droplets. These droplets then enter the engine's combustion chamber, where they face high temperatures and pressure. The evaporation of fuel droplets is impacted by a few factors, including the size and distribution of the droplets, combustion chamber temperature and pressure, the chemical composition of the fuel, and the presence of other substances, such as air or vapor. To optimize the evaporation of fuel droplets, a few strategies can be employed, including the use of fuel injectors with specific design features, the addition of special additives to the fuel, and the control of engine parameters, such as temperature, pressure, and air-fuel ratio [29]. For combustion of fuel droplets refers to the process by which the vaporized fuel in an engine's combustion chamber reacts with air to release energy. This reaction is commonly referred to as burning. The combustion of fuel droplets is a critical part of the internal combustion engine, as it provides the energy needed to power the engine. The combustion process is initiated when the vaporized fuel and air are mixed in the proper ratio to support ignition. This mixture is then subjected to a spark or compression, which triggers the reaction [30]. The exhaust effect refers to the phenomenon in which the flow of exhaust gases from an engine or other combustion device affects the behavior of the combustion process. This effect can cause impact on the performance and efficiency of the engine or device. The exhaust effect can result from a few factors, including the velocity, temperature, and pressure of the exhaust gases, as well as their chemical composition. For example, the pressure and velocity of the exhaust gases can create a pressure differential that influences the flow of air and fuel into the combustion chamber, leading to changes in the combustion process [31]. The exhaust effect can also play a role in the formation of emissions and pollutants, as the pressure and temperature conditions in the exhaust stream can influence the conversion of harmful substances into other forms. To mitigate the negative effects of the exhaust effect, a variety of strategies can be employed, including the use of exhaust gas recirculation systems, catalytic converters, and other after-treatment technologies. In summary, the exhaust effect refers to the phenomenon in which the flow of exhaust gases from an engine or other combustion device affects the behavior of the combustion process. Strategies can be employed to mitigate the negative effects of the exhaust effect [32].

2. Modeling evaporation of fuel droplets

Modeling the evaporation of droplets under different conditions is described in [33] in detail. Numerical modeling of droplet evaporation involves the conservation equations of mass, energy and momentum. Apart from these equations droplet evaporation involve the heat transfer and mass diffusion phenomenon. When heat is transferred to the droplet then the mass of droplet is lost. Turbulence is also a key part of the evaporation modeling at elevated temperature and pressure conditions. Evaporation of fuel droplets for diesel and biodiesel fuels is numerically modeled and verified in [34–37] and model was applied in Ansys Fluent. The numerical model equations are given in detail as follows.

2.1 Continuity equation

The general continuity equation is given in the following form.

$$\frac{\partial \rho}{\partial t} + \nabla.\left(\rho \vec{v}\right) = S_m \tag{1}$$

Mass of the fuel droplet is conserved by the Eq. (1). The term on the right side of the equation is the source term. This term implies the mass of fuel droplets added into the engine cylinder from the atomization of fuel in form of spray.

2.2 Momentum equation

Conservation of momentum equation is applied to account for the momentum associated with the droplet. When the liquid fuel is injected into the combustion chamber, it has some velocity by which it travels along the piston head. As the liquid fuel is sprayed, there are certain forces, which are associated with the fuel droplets. These forces include the inertial forces, body forces, gravitational forces and drag forces. All these forces are taken into account by applying the momentum equation, which is given as (2).

$$\frac{\partial}{\partial t}\left(\rho\vec{v}\right)+\nabla\left(\rho\vec{v}\vec{v}\right)=-\nabla p+\nabla(\bar{\bar{\tau}})+\rho\vec{g}+\vec{F} \quad (2)$$

$$\bar{\bar{\tau}}=\mu\left[\left(\nabla\vec{v}+\nabla\vec{v}^{T}\right)-\frac{2}{3}\nabla.\vec{v}I\right] \quad (3)$$

2.3 Energy equation

For droplet evaporation following energy is solved in the Ansys Fluent. Fuel droplet enters in the hot environment after the compression stroke in the engine cylinder. When the droplet is injected in form of fine spray at high pressure in the hot environment, it absorbs the heat present in the engine cylinder. The amount of energy transfer is governed by the Eq. (4), which involves the latent heat and diffusion heat flux parameters.

$$\frac{\partial}{\partial t}(\rho E)+\nabla\left(\vec{v}(\rho E+p)\right)=\nabla\left(k_{eff}\nabla T-\sum_{j}h_j\vec{J}_j+\left(\bar{\bar{\tau}}_{eff}.\vec{v}\right)\right)+S_h\frac{1}{2} \quad (4)$$

$$E=h-\frac{p}{\rho}+\frac{v^2}{2} \quad (5)$$

$$h=\sum_{j}Y_jh_j \quad (6)$$

$$h_j=\int_{T_{ref}}^{T}c_{p,j}dT \quad (7)$$

2.4 Species transport equation

This equation associated with the diffusion of mass of liquid fuel into the continuous phase. This is a convection-diffusion equation used to solve for the j species. This equation is solved to predict the local mass fraction of the species in the combustion chamber. It involves the binary diffusion coefficient of the fuel droplets and turbulent Schmidt number, which are used to govern the local mass fraction of liquid fuel specie in the continuous phase, which is compressed air.

$$\frac{\partial}{\partial t}(\rho Y_j)+\nabla.\left(\rho\vec{v}Y_j\right)=-\nabla.\vec{J}_j+R_j+S_j \quad (8)$$

$$\vec{J}_j=-\left(\rho D_{j,m}+\frac{\mu_t}{Sc_t}\right)\nabla Y_j-D_{T,j}\frac{\nabla T}{T} \quad (9)$$

$$Sc_t = \frac{\mu_t}{\rho D_t} \tag{10}$$

2.5 Particle force balance equation

Inertial force on the fuel droplet is balanced by the particle drag force, gravitational force and external body force and is given by the following equation. Calculation of drag force involves the coefficient of drag and relative Reynolds number of the droplet. Apart from these, droplet diameter and density of liquid fuel are required to calculate the drag force on the droplet.

$$\frac{dv_p}{dt} = F_d(v - v_p) + \left(\rho_p - p\right)\frac{g_x}{\rho_p} + F_x \tag{11}$$

$$F_d = \frac{18\mu}{\rho_p {d_p}^2} \cdot \frac{C_d \,\mathrm{Re}}{24} \tag{12}$$

$$\mathrm{Re} = \frac{\rho d_p}{\mu}\left|u_p - u\right| \tag{13}$$

2.6 Heat transfer equation

Heat is transferred to the droplet by following equation. When droplet is injected into the combustion chamber, the droplet absorbs the heat present in the cylinder due to high compression. The heat transfer equation involves the mass of droplet particle, temperature of droplet, ambient temperature of engine cylinder, surface area of droplet and latent heat of vaporization of droplet.

$$m_p c_p \frac{dT_p}{dt} = hA_p(T_\infty - T_p) + \frac{dm_p}{dt} h_{fg} \tag{14}$$

2.7 Mass transfer equation

When heat is transferred to the droplet, the mass transfer takes place. Mass of droplet is lost through the diffusion. Mass of droplet is decreased according to the following equation

$$m_p(t + \Delta t) = m_p(t) - N_j A_p M_{\omega,\mathrm{j}} \Delta t \tag{15}$$

$$N_j = k_c\left(C_{\mathrm{j},s} - C_{\mathrm{j},\infty}\right) \tag{16}$$

$$Nu_{AB} = \frac{k_c d_p}{D_{j,m}} \tag{17}$$

2.8 Turbulence model equations

Turbulence is associated with the fuel droplet during its evaporation and break up regimes in the engine cylinder. The relative Reynolds number given by Eq. (13) predicts turbulence around the fuel droplets. To take into account the turbulence effects around the fuel droplet a turbulence model is mandatory with the evaporation modeling. There are three turbulence models than be applied for the droplet evaporation. These include the Direct Numerical Simulation (DNS), Large Eddy Simulation

(LES) and Reynolds Average Navier-Stokes (RANS) model. Transport equations for the Realizable k-e model are given below. Each model has own advantages and limitations. DNS and LES are computationally much expensive than the RANS turbulence model. There are further three classifications in RANS turbulence. Realizable K-epsilon model can be used for the droplet evaporation modeling as it is computationally inexpensive and gives the acceptable results.

$$\frac{\partial}{\partial t}(\rho k) + \frac{\partial}{\partial X_j}(\rho k u_j) = \frac{\partial}{\partial X_j}\left[\left(\mu + \frac{\mu_t}{\sigma_k}\right)\frac{\partial k}{\partial X_j}\right] + G_k + G_b - \rho\varepsilon - Y_M + S_k \tag{18}$$

$$\begin{aligned}\frac{\partial}{\partial t}(\rho\varepsilon) + \frac{\partial}{\partial X_j}(\rho\varepsilon u_j) = \frac{\partial}{\partial X_j}\left[\left(\mu + \frac{\mu_t}{\sigma_\varepsilon}\right)\frac{\partial \varepsilon}{\partial X_j}\right] + \\ \rho C_1 S\varepsilon - \rho C_2\frac{\varepsilon^2}{k + \sqrt{v\varepsilon}} + C_{1\varepsilon}\frac{\varepsilon}{k}C_3 G_b + S_\varepsilon\end{aligned} \tag{19}$$

$$C_1 = \max\left[0.43, \frac{\eta}{\eta + 5}\right], \left(\eta = S\frac{k}{\varepsilon}\right), \left(S = \sqrt{2S_{ij}S_{ij}}\right) \tag{20}$$

$$\mu_t = \rho C_\mu \frac{k^2}{\varepsilon} \tag{21}$$

$$C_\mu = \frac{1}{A_o + A_s\frac{kU^*}{\varepsilon}} \tag{22}$$

3. Properties of fuels

Various properties play important roles in the evaporation of fuel droplets in the engine cylinder. These include the density, viscosity, latent heat of vaporization, boiling point, vaporization temperature and volatile component fraction. These properties are given in **Table 1** from [34].

Droplet properties	Diesel	Thumba biodiesel
Droplet surface tension (N/m)	0.02521	0.02715
Vaporization temperature (K)	341	341
Specific heat capacity (J/kg K)	2090	1774
Volatile component fraction (%)	100	100
Thermal conductivity (W/m K)	0.149	0.158
Density (kg/m^3)	835	880
Boiling point (K)	447	691.69
Saturation vapor pressure (Pascal)	1329	1329
Latent heat (J/kg)	277,000	229,327
Binary diffusivity (m^2 /s)	3.79×10^{-6}	7.42×10^{-6}
Viscosity (kg/m s)	0.004	0.004664

Table 1.
Properties of fuels.

4. Results and discussion

Evaporation of diesel fuel and biofuel droplets is modeled in [34] using the computational fluid dynamics techniques. In this study, three different ambient temperatures of 623, 823 and 973 K are taken. The diameter of droplet is taken as 20 and 25 μm. There are two different fuels. First one is n-decane diesel fuel and second is Thumba bio-diesel. Numerical model is validated with the most accurate vaporization experiments of Chauveau [38] and the model of Abramzon and Sirignano (AS-1989) [39]. Turbulence effects present in the engine cylinder are also taken into account by applying the K-epsilon realizable turbulence models. Results include the droplet lifetime, increase in temperature of droplet and reduction in velocity of droplet. Results indicate that droplets having larger diameter take more time to evaporate while the small droplet need short time for complete evaporation. It is also observed that at higher temperatures biofuels evaporates faster than the conventional diesel. Complete evaporation of droplets results in the higher thermal efficiency of internal combustion engine. **Figure 1** shows the one-sixth sector of piston geometry, which is under consideration. The geometry of piston is divided into 6 equal parts for the simplicity of analysis and inexpensive computations.

In **Figure 2**, droplet decay profiles of 20-μm fuel droplets of diesel and biodiesel are shown at various ambient temperatures. Three different temperatures of 973, 823 and 623 K are taken for analysis purpose. It is observed that biodiesel droplets have shorter life span in the engine cylinder as compared to the conventional diesel fuel droplets. As the temperature of cylinder increased after the compression stroke, the droplet lifetime is decreased significantly as evident from **Figure 2**.

In **Figure 3**, droplet decay profiles for 25-μm fuel droplets are shown. The trend in the profiles is same as in the **Figure 2**. Biodiesel is evaporated earlier than the diesel fuel. As the ambient temperature increased, evaporation rate is also increased.

In **Figure 4**, temperature profiles of 20-μm droplets of diesel and biodiesel are shown.

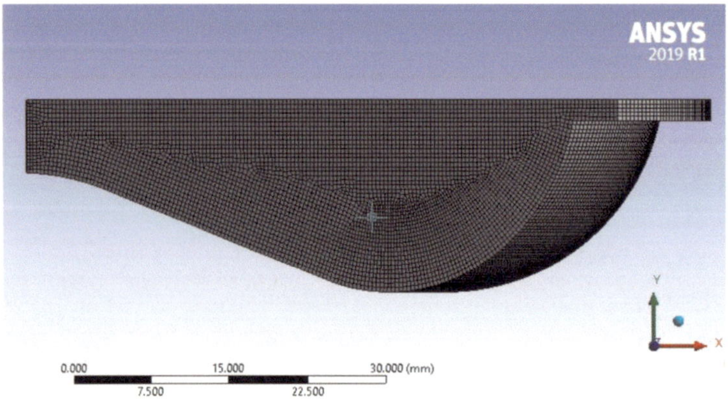

Figure 1.
Mesh of one-sixth sector of piston geometry.

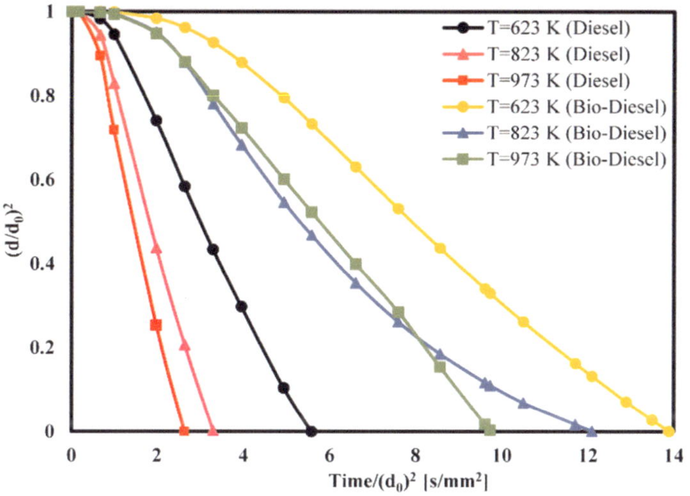

Figure 2.
20-μm droplet profiles at different ambient temperatures, reprinted from [34] CC BY 4.0.

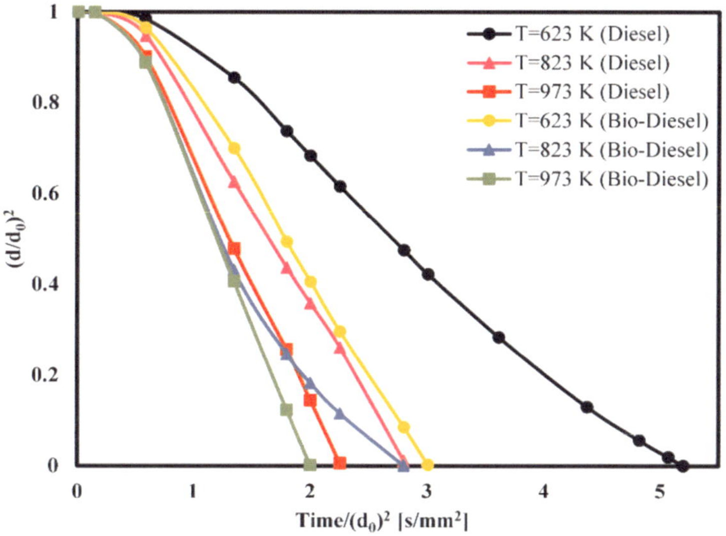

Figure 3.
25-μm droplet profiles at different ambient temperatures, reprinted from [34] CC BY 4.0.

It is observed that the trend in the temperature profiles matches with that of decay in diameter profiles. However, the temperature achieved by the biodiesel droplets is more than the diesel fuel droplets. It is due the fact that the boiling point of the biodiesel is greater than the diesel fuel droplets. The heat up period of the biodiesel gives maximum temperature to the fuel droplets before complete evaporation.

Figure 5 shows the velocity profiles of 20-μm fuel droplets. Residence time of biodiesel droplets is less than the diesel droplets as evident in **Figure 5**. The lowest velocity is achieved by the biodiesel droplet at 873 K. Diesel fuel droplet has the largest residence time at lower temperature of 623 K.

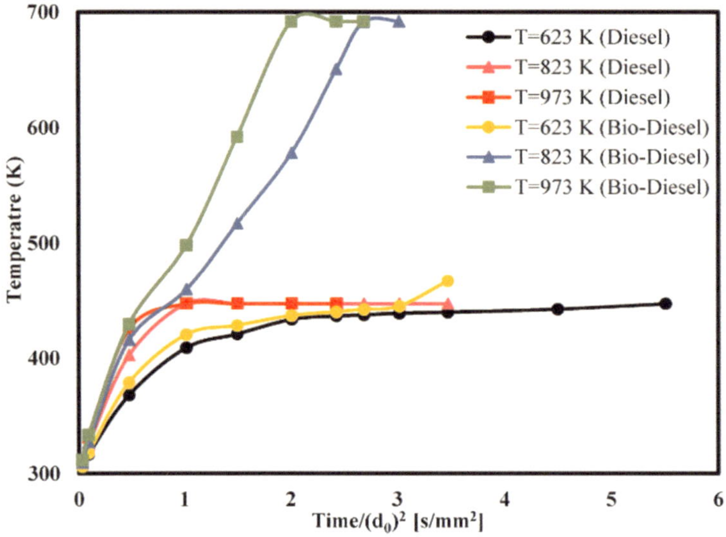

Figure 4.
Temperature profiles of 20-μm fuel droplet at different ambient temperatures, reprinted from [34] CC BY 4.0.

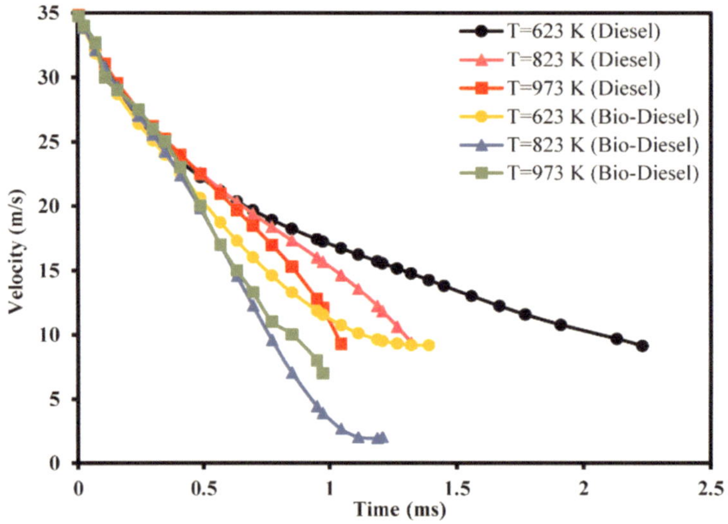

Figure 5.
Velocity profiles of 20-μm fuel droplet at different ambient temperatures, reprinted from [34] CC BY 4.0.

5. Conclusion

In this chapter, an overview of renewable energy resources in terms of biofuels has been described. A comparison of evaporation of two different fuels is drawn. Several in cylinder processes are described in detail. There are different ways to model the evaporation of liquid fuel droplets. Evaporation results can be obtained by experiments using the high-speed cameras while numerical modeling of evaporation is also a known phenomenon by using the conservation equations of mass, momentum and energy and other droplet equations. These numerical equations can be solved using the commercial soft wares e.g. Ansys, open foam, KIVA and Star CD. Evaporation of diesel and biodiesel fuels is observed at different diameters and different ambient temperatures. It is concluded that at higher temperature biofuel droplets show the fast evaporation rate than the conventional diesel fuel. Droplets with large size take more time to evaporate while the droplets with smaller diameters take lower time to evaporation completely. Thus, biodiesel is a potential alternative fuel for the transportation sector as a replacement of conventional diesel with low emission rate.

Conflict of interest

The authors declare no conflict of interest.

Nomenclature

A_p	surface area of particle (m^2)
C_d	drag coefficient
C_p	droplet heat capacity (J/kg K)
$C_{i,\infty}$	vapor concentration in bulk gas (kg mol/)
$C_{i,s}$	vapor concentration at droplet surface (kg mol/m^3)
$D_{j,m}$	mass diffusion coefficient of species j (m^2/s)
d_p	droplet diameter (m)
$D_{T,m}$	thermal diffusion coefficient of species j (m^2/s)
F	external body force (N/m^2)
h	convective heat transfer (W/m^2 K)
h_{fg}	latent heat of vaporization (J/kg)
I	unit tensor
j_j	diffusion flux of species j (kg/s m^2)
k_{eff}	effective thermal conductivity (W/m K)
k_c	mass transfer coefficient (m/s)
m_p	mass of particle (kg)
$M_{w,j}$	molecular weight of species j (kg/kg mol)
N_j	molar flux of vapor (kgmol/m^2 s)
p	static pressure (N/m^2)
Re	relative Reynolds number
R_j	net rate of production of species j
S_j	rate of creation by addition from disperse phase
Sc_t	turbulent Schmidt number
S_m	source term
S_h	heat of chemical reaction
T_∞	temperature of continuous phase (K)
v_p	velocity of droplet (m/s)
v	velocity of air (m/s)
Y_j	mass fraction of species j
ρ	density of air (m^3/kg)
ρ_p	density of particle (m^3/kg)
τ	stress tensor (N/m^2)
μ	molecular viscosity (kg/ms)
μ_t	turbulent viscosity (m^2/s)

Author details

Ali Raza*, Zunaira-Tu-Zehra, Muhammad Khurram,
Muhammad Ahsan Pervaiz Khan, Asif Durez and Liaquat Ali Khan
National University of Technology (NUTECH), Islamabad, Pakistan

*Address all correspondence to: ali.raza@nutech.edu.pk

References

[1] Sorensen B. Renewable energy: A technical overview. Energy Policy. 1991; **19**(4):386-391

[2] IPCC. Summary for policymakers. In: Edenhofer O, Pichs-Madruga R, Sokona Y, Seyboth K, Matschoss P, Kadner S, Zwickel T, Eickemeier P, Hansen G, Schlömer S, Stechow, C.v., editors. IPCC Special Report on Renewable Energy Sources and Climate Change Mitigation. Cambridge, United Kingdom and New York, NY, USA: Cambridge University Press; 2011

[3] Kannan N, Vakeesan D. Solar energy for future world: A review. Renewable and Sustainable Energy Reviews. 2016; **62**:1092-1105

[4] Herbert GMJ et al. A review of wind energy technologies. Renewable and Sustainable Energy Reviews. 2007;**11**(6): 1117-1145

[5] Rehman S, Al-Hadhrami LM, Alam MM. Pumped hydro energy storage system: A technological review. Renewable and Sustainable Energy Reviews. 2015;**44**:586-598

[6] Barbier E. Geothermal energy technology and current status: An overview. Renewable and Sustainable Energy Reviews. 2002;**6**(1–2):3-65

[7] Zia UUR et al. Dependence of bio energy production on chemical composition and crop phenology of biomass feedstock. In: 2019 International Conference on Electrical, Communication, and Computer Engineering (ICECCE). Swat, Pakistan: IEEE; 24-25 July 2019

[8] Khan N et al. Review of ocean tidal, wave and thermal energy technologies. Renewable and Sustainable Energy Reviews. 2017;**72**:590-604

[9] Demirbas MF, Balat M, Balat H. Biowastes-to-biofuels. Energy Conversion and Management. 2011; **52**(4):1815-1828

[10] Shahzad U. The need for renewable energy sources. Energy. 2012;**2**:16-18

[11] Heal G. The Economics of Renewable Energy. Medford, MA: Tufts University; 2009

[12] Qazi A et al. Towards sustainable energy: A systematic review of renewable energy sources, technologies, and public opinions. IEEE Access. 2019; **7**:63837-63851

[13] Saleem H, Khan MB, Shabbir MS, Khan GY, Usman M. Nexus between non-renewable energy production, CO_2 emissions, and healthcare spending in OECD economies. Environmental Science and Pollution Research. 2022 Jul;**29**(31): 47286-47297. DOI: 10.1007/s11356-021-18131-9. Epub 2022 Feb 18. PMID: 35179687

[14] Prothero DR. Evolution: What the Fossils Say and Why it Matters. New York, NY: Columbia University Press; 2007

[15] Veloz L. Importance of Fossils scienc ing.com. 2023. Available from: https:// sciencing.com/importance-fossils-2470. html

[16] Available from: https://byjus.com/ physics/types-of-fossils/ [Accessed: 16 January 2023]

[17] Archer D et al. Atmospheric lifetime of fossil fuel carbon dioxide. Annual Review of Earth and Planetary Sciences. 2009;**37**:1-18

[18] Bromley RG. Trace Fossils: Biology, Taxonomy and Applications. Milton

Park, Abingdon, Oxfordshire: Routledge; 2012

[19] Schopf JM. Modes of fossil preservation. Review of Palaeobotany and Palynology. 1975;**20**(1–2):27-53

[20] Chmielewski AG. Environmental Effects of Fossil Fuel Combustion. Interaction: Energy/Environment; 1999

[21] Maciejczyk P, Chen L-C, Thurston G. The role of fossil fuel combustion metals in PM2. 5 air pollution health associations. Atmosphere. 2021;**12**(9):1086

[22] Rodionova MV, Poudyal RS, Tiwari I, Voloshin RA, Zharmukhamedov SK, Nam HG, et al. Biofuel production: Challenges and opportunities. International Journal of Hydrogen Energy. 2017;**42**(12): 8450-8461

[23] Sims RE, Mabee W, Saddler JN, Taylor M. An overview of second-generation biofuel technologies. Bioresource Technology. 2010;**101**(6): 1570-1580

[24] Godin B, Lamaudière S, Agneessens R, Schmit T, Goffart JP, Stilmant D, et al. Chemical composition and biofuel potentials of a wide diversity of plant biomasses. Energy & Fuels. 2013;**27**(5):2588-2598

[25] Demirbas A. Biofuels sources, biofuel policy, biofuel economy and global biofuel projections. Energy Conversion and Management. 2008; **49**(8):2106-2116

[26] Sou A, Maulana MI, Isozaki K, Hosokawa S, Tomiyama A. Effects of nozzle geometry on cavitation in nozzles of pressure atomizers. Journal of Fluid Science and Technology. 2008;**3**(5): 622-632

[27] Lebas R, Menard T, Beau PA, Berlemont A, Demoulin FX. Numerical simulation of primary break-up and atomization: DNS and modelling study. International Journal of Multiphase Flow. 2009;**35**(3):247-260

[28] Lasheras JC, Villermaux E, Hopfinger EJ. Break-up and atomization of a round water jet by a high-speed annular air jet. Journal of Fluid Mechanics. 1998;**357**:351-379

[29] Sazhin SS. Modelling of fuel droplet heating and evaporation: Recent results and unsolved problems. Fuel. 2017;**196**: 69-101

[30] Kumagai S, Sakai T, Okajima S. Combustion of free fuel droplets in a freely falling chamber. In: Symposium (International) on Combustion. Vol. 13(1). USA: Elsevier; 1971. pp. 779-785

[31] Ma Y, Zhu M, Zhang D. The effect of a homogeneous combustion catalyst on exhaust emissions from a single cylinder diesel engine. Applied Energy. 2013;**102**: 556-562

[32] Hussain J, Palaniradja K, Alagumurthi N, Manimaran R. Retracted: Effect of Exhaust Gas Recirculation (EGR) on Performance and Emission Characteristics of a Three Cylinder Direct Injection Compression Ignition Engine. CA, USA: Rosemead; 2012

[33] Sazhin S. Droplets and Sprays. Vol. 345. London: Springer; 2014

[34] Raza A et al. Investigation on the characteristics of biodiesel droplets in the engine cylinder. Energies. 2020; **13**(14):3637

[35] Raza A et al. Numerical modelling of diesel fuel multiphase evaporation in

heavy duty diesel engine. In: Proceedings of the 21st Australasian Fluid Mechanics Conference (AFMC), Adelaide, Australia. 2018

[36] Raza A et al. Numerical investigation of evaporation modelling for different diesel fuels at high temperature and pressure in diesel engine. Journal of Engineering Research. ICEPE Special Issue. 2022:1-15

[37] Raza A et al. Numerical study of evaporation modelling for different fuels at high operating conditions in a diesel engine. Engineering Proceedings. 2021; **12**(1):8

[38] Chauveau C, Halter F, Lalonde A, Gökalp I. An experimental study on the droplet vaporization: Effects of heat conduction through the support fiber. In: Proceedings of the 22nd Annual Conference on Liquid Atomization and Spray Systems (ILASS Europe 2008), Como Lake, Italy, 8–10 September 2008.

[39] Abramzon B, Sirignano W. Droplet vaporization model for spray combustion calculations. International Journal of Heat and Mass Transfer. 1989; **32**:1605-1618

Section 2

Applications of Exergy

Chapter 5

Computational Simulation of Heat Transfer in a Dip Shrink Tank Using Two Different Arrangements of Electrical Resistances

José Luis Velázquez Ortega and Aldo Gómez López

Abstract

Biohazard recontamination of food can occur in a meat processing plant during slicing, portioning, or racking. Subsequently, to protect them from external agents, not allow the loss of moisture contained in the product, and preserve its safety, they undergo a shrinking process; which consists of submerging in a tank with hot water at an approximate temperature of 87°C, for a certain time the food that has been wrapped with a heat shrink plastic, making it shrink. In this work, the behavior of heat transfer in a non-commercial shrink tank, built with two different arrangements of electrical resistances for water heating, is investigated. The study was carried out through numerical simulations with the implicit method of alternating directions (ADI). The results obtained from the heating times with their respective temperature distributions show that the arrangement with four resistances is the most efficient for the process of heating the water in the shrink tank, achieving a homogeneous temperature of 87°C, in times less than 9 minutes with a heat flux of q = 24.48 W. The validation of the simulations will be carried out in a subsequent work with experimental tests carried out in the shrink tank.

Keywords: convective heat transfer, dip shrink tank, temperature distribution, stream function, vorticity

1. Introduction

The packaging of a food product consists of placing a material known as packaging, which completely covers the object, which can generally have two functions, the main and most important is to protect it and the second is the visual impact for commercial purposes.

Many foods require a container that guarantees their conservation, taking into account their different properties (liquid, solid, gel, pH, among others) and composition (proteins, lipids, vitamins, etc.). For this purpose, various materials have been used, which could have harmful properties for food, if the correct material and properties are not chosen.

The heat shrink process consists of immersing food previously covered in heat shrink plastic (in general, the material used for this process is polyethylene), in a tank with hot water, for a certain time with the intention of protecting it from external agents [1].

In many meat derivatives processing plants, a problem associated with product recontamination after its primary heat treatment has been detected, mainly associated with the use of a cooking sleeve that is changed for marketing the product, as well as the removal of that sleeve to distribute the product sliced or in portions and vacuum packed. Subsequently, these presentations are subjected to the shrinking process, preceded by some other procedure that allows control of recontamination, for example, the treatment by High Hydrostatic Pressures (HPP) [2].

For this reason, it is essential to obtain the normal operating conditions of temperature and time in the shrink tank, which allow increasing and sustaining these variables sufficiently to control the biological hazards that are identified.

In this work, the heat transfer was investigated in a shrink tank built in the facilities of the Faculty of Higher Studies Cuautitlán—UNAM, with two different arrangements of electrical resistances for water heating, in order to select an arrangement of resistances to ensure adequate heat transfer within the tank.

The behavior of heat transport that was studied is convection, which is considered an improved or modified form of conduction, in which a massive movement of the medium is also present [3]. There are two types of mechanisms for convection, free and forced. In this investigation we worked with free convection.

To model the shrink tub system, a rectangular cavity open at the top in two dimensions was considered, with two different configurations of electrical resistances, which serve as heat sources.

The resulting system of equations was discretized and a finite difference scheme was also used, which was solved using the Method of Alternate Directions Implicit (ADI), obtaining as a result the profiles of the stream function φ and vorticity ω.

2. Theory and models

2.1 Governing equations

Conduction, convection, and radiation are identified as ways to transfer heat. The mechanism of conductive heat transfer can be appreciated by heating a material with a heat source, as shown in **Figure 1**.

In the previous figure, there is a higher temperature on the left side of the material due to the heat source, and on the right side a lower temperature; Therefore, this temperature difference will result in heat transport by conduction in the material.

The heat flow is proportional to the area and the temperature difference, and can be quantified by means of Fourier's first law.

$$Q = kA\left(-\frac{dT}{dx}\right) \tag{1}$$

In Eq. (1), k is the thermal conductivity and depends on the material, A is the cross-sectional area and (dT/dx) is known as the temperature gradient, and the negative sign indicates that the heat flux is in the opposite direction of the temperature gradient [4].

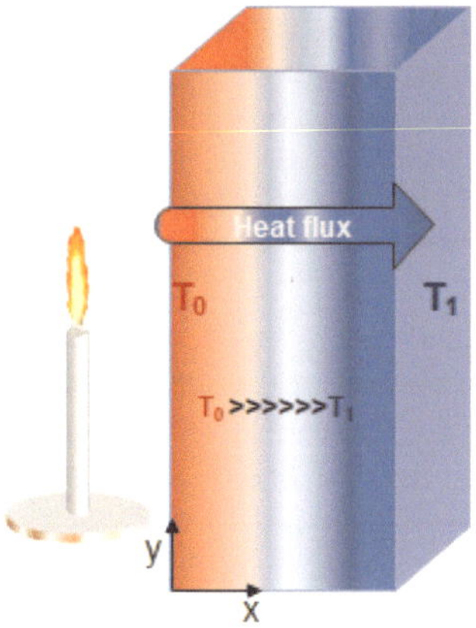

Figure 1.
Heating a material.

If one wanted to quantify the heat flux, Eq. (1) becomes

$$q = k\left(-\frac{dT}{dx}\right) \tag{2}$$

For the case of transient heat transport, Joseph Fourier developed a second law, which considers the variation of temperature with respect to time and is given by

$$\rho C_p \frac{\partial T}{\partial t} = k\nabla^2 T \tag{3}$$

In Eq. (3), ρ corresponds to the density and C_p to the heat capacity at constant pressure.

Regarding the mechanism of heat transport by convection, it can be seen by heating a metal container containing water, as shown in **Figure 2**.

In the previous figure, it can be seen that the fire produced by the combustion of the gas in the stove heats the container and this, in turn, heats the water. At a certain time, the fluid at the bottom of the container will have a lower density compared to the fluid near the surface, this is due to its thermal expansion. Therefore, the liquid at the bottom tends to rise and the liquid on the surface will go down, resulting in a heat

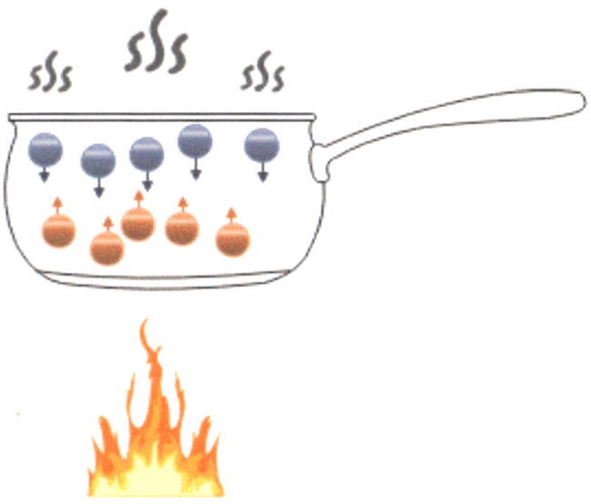

Figure 2.
Heating of a container containing water.

transfer when mixed. Therefore, the phenomenon of convection occurs when mixing relatively hot portions of fluid with cold ones. The equation with which the heat flow by convection is quantified is Newton's law of cooling.

$$Q = -hA\Delta T \tag{4}$$

In Eq. (4), h is the convective coefficient, A is the area, and ΔT the temperature difference [5].

As for radiation, it refers to the radiant energy from a source to a receiver; in which part of the energy is absorbed by the receiver and part reflected by it. Boltzmann established an equation for the flow of heat by radiation

$$Q = \varepsilon\sigma AT^4 \tag{5}$$

In Eq. (5), σ is the dimensionless Boltzmann constant and ε the emissivity [3].

For the case in which internal heat generation and convective heat transport are present, Eq. (3) takes the following form

$$\rho C_p \frac{\partial T}{\partial t} = k\nabla^2 T + G - \nabla \cdot \rho C_p (T - T_{ref})\vec{v} \tag{6}$$

Eq. (6) contemplates the heat flow in a transitory state, the conduction heat mechanism ($k\nabla^2 T$), the internal heat generation (G) due to electrical resistances and convection $\nabla \cdot \rho C_p (T - T_{ref})\vec{v}$, being the velocity and T_{ref}, a reference temperature.

In the case of fluid mechanics, there are two equations that are essential to characterize the flow of fluids, these are the continuity equation and the Navier-Stokes equation, which are expressed in a vector formulation as follows [6]:

$$\frac{\partial \rho}{\partial t} = \nabla \cdot \rho\vec{v} \tag{7}$$

Eq. (7) corresponds to the continuity equation, in which ρ is the density and $\vec{v}$, is the velocity.

$$\rho\left[\frac{\partial \vec{v}}{\partial t} + \vec{v} \cdot \nabla\vec{v}\right] = \mu\nabla^2\vec{v} - \nabla P + \rho\vec{g} \tag{8}$$

Eq. (8) is the well-known Navier-Stokes equation, the term μ denotes the viscosity, P is the pressure, and $\vec{g}$ is the acceleration due to gravity.

2.2 Materials

The shrink tank used for the simulations can be seen in **Figure 3**, which was built with 304 stainless steel material for the parts that are in contact with the meat product. The dimensions of the tank are 0.728 m × 0.420 m.

2.3 Modeling

For the numerical modeling, a two-dimensional rectangular cavity open at the top was considered, with the following configurations for the placement of electrical resistances as heat sources in red, as shown in **Figure 4**.

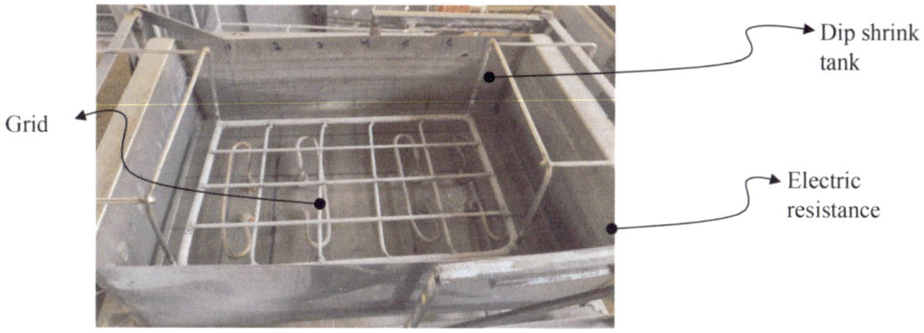

Figure 3.
Shrink tank.

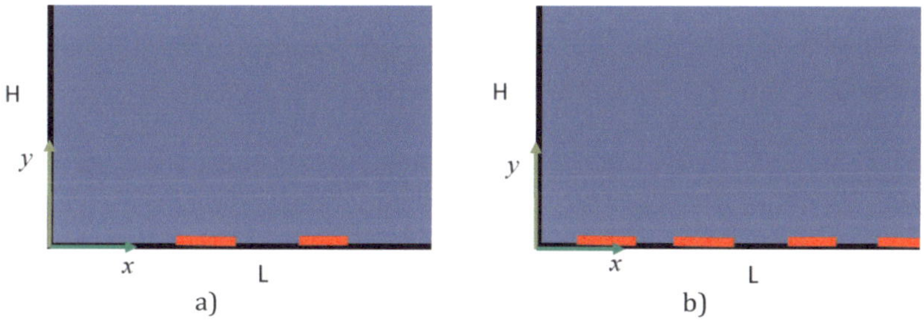

Figure 4.
(a) Arrangement of two resistors and (b) arrangement of four resistors for the rectangular cavity.

The cavity has a height H and a length L, a relationship between both magnitudes was considered as L = 3H. In the same way, the heat sources that are indicated with red color in **Figure 4**, were considered as 1/9 of L.

For the first configuration, the distance from the walls to the heat sources was estimated as 1/3 of L. For the second configuration, the distance from the wall to the heat sources is 1/9 of L and the distance between the different sources of heat is 1/9 of L.

All heat sources are considered equal and constantly emit heat.

For modeling, the temperatures (T_0) of the cavity walls and the upper zone were considered constant. The working fluid is water and no induced flow or pressure gradient is considered, in addition the upper zone is considered open to the environment, and it is assumed that there is no fluid exchange with the external environment.

It is considered that, due to temperature gradients, the fluid experiences density changes small enough to support the hypothesis of an incompressible fluid but large enough to produce a convection phenomenon, for which the Boussinesq approximation is taken.

According to the above, the conservation equations can be established as follows:

- The equation of conservation of mass for an incompressible fluid is obtained from Eq. (7) and remains as

$$\nabla \cdot \vec{v} = 0 \tag{9}$$

- The momentum conservation equation with the convection term is

$$\frac{\partial \vec{v}}{\partial t} + \vec{v} \cdot \left(\nabla \vec{v}\right) = -\frac{1}{\rho_0} \nabla \left(P - \rho_0 \vec{g}\right) + \nabla^2 \vec{v} + \overline{g}\beta(T - T_0) \tag{10}$$

In the above equation, ρ is the initial density of the fluid, β is the coefficient of thermal expansion, $\vec{g}$ is the gravity vector (which only points in the y direction), t is time, ν is the kinematic viscosity, and T_0 is the external temperature.

- Equation of conservation of energy

$$\frac{\partial T}{\partial t} + \vec{v} \cdot (\nabla T) = \alpha \nabla^2 T \tag{11}$$

In the above equation, the relationship $k/_{\rho C_p}$ that appears in Eq. (3) represents the diffusion coefficient α.

For the system being studied, it is considered that the fluid is static and with external temperature when t = 0. Similarly, the system is considered only in two dimensions, therefore $\vec{v} = (u, v, 0)$, where "u" and "v" are the velocity components in the "x" and "y" direction respectively and are functions of time.

With the above, they were established as boundary conditions

$$u(t, x, 0) = 0, u\left(t, 0, y\right) = 0, u\left(t, L, y\right) = 0, \frac{\partial u}{\partial y}(t, x, H) = 0 \tag{12}$$

$$v(t, x, 0) = 0, v\left(t, 0, y\right) = 0, v\left(t, L, y\right) = 0, v(t, x, H) = 0 \tag{13}$$

For the temperature we have

$$T(t, x, 0) = T_0, T\left(t, 0, y\right) = T_0, T\left(t, L, y\right) = T_0, T(t, x, H) = T_0 \tag{14}$$

Except for the points where the heat sources are located, as already mentioned above.

The previous system was expressed in dimensionless form considering the following characteristic variables

$$x^* = \frac{x}{H}, y^* = \frac{y}{H}, T = \frac{T - T_0}{T_H - T_o}, t^* = \frac{\alpha}{H^2} t \tag{15}$$

$$p^* = \frac{H^2}{\rho_0 \alpha^2} p, u^* = \frac{H}{\alpha} u, v^2 = \frac{H}{\alpha} v \tag{16}$$

The constant T_H is defined from the heat source as

$$T_H = \frac{qH}{K} + T_0 \tag{17}$$

By introducing the previous dimensionless variables to the system, we obtain

$$\frac{\partial u^*}{\partial x^*} + \frac{\partial v^*}{\partial y^*} = 0 \tag{18}$$

$$\frac{Du^*}{Dt^*} = -\frac{\partial P^*}{\partial x^*} + Pr\left(\frac{\partial^2}{\partial x^{*2}}u^* + \frac{\partial^2}{\partial y^{*2}}v^*\right) \tag{19}$$

$$\frac{Dv^*}{Dt^*} = -\frac{\partial P^*}{\partial y^*} + Pr\left(\frac{\partial^2}{\partial x^{*2}}u^* + \frac{\partial^2}{\partial y^{*2}}v^*\right) + RaPrT^* \tag{20}$$

$$\frac{DT^*}{Dt^*} = \frac{\partial^2 T^*}{\partial x^{*2}} + \frac{\partial^2 T^*}{\partial y^{*2}} \tag{21}$$

where Pr is the Prandtl number and Ra is the Rayleigh number, which are defined as

$$Pr = \frac{\nu}{\alpha}, Ra = \frac{\beta g L^3 (T_H - T_0)}{\nu\alpha} \tag{22}$$

In Eq. (22), the operator $D/_{Dt}$ represents the material derivative that is defined as $\frac{DA}{Dt} = \frac{\partial A}{\partial t} + \vec{v} \cdot (\nabla A)$. The term P* is defined as $P^* = p^* + gy^* L^3/\alpha^2$.

For simplicity, from now on the use of (*) to indicate dimensionless variables will be omitted, assuming that all equations are in dimensionless form.

In order to reduce the system, the stream function formulation, φ, and vorticity, ω, are applied, which are defined as [7]

$$u = \frac{\partial\varphi}{\partial y}, v = -\frac{\partial\varphi}{\partial x}, \omega = -\frac{\partial u}{\partial y} + \frac{\partial v}{\partial x} \tag{23}$$

so the system is rewritten as follows

$$\frac{\partial^2\varphi}{\partial x^2} + \frac{\partial^2\varphi}{\partial y^2} = -\omega \tag{24}$$

$$\frac{D\omega}{Dt} = Pr\left(\frac{\partial^2\omega}{\partial x^2} + \frac{\partial^2\omega}{\partial y^2}\right) + RaPr\frac{\partial T}{\partial x} \tag{25}$$

$$\frac{DT}{Dt} = \frac{\partial^2 T}{\partial x^2} + \frac{\partial^2 T}{\partial y^2} \tag{26}$$

With this new formulation, the new initial conditions are

$$\varphi(0, x, y) = 0, \omega(0, x, y) = 0, T(0, x, y) = 0 \tag{27}$$

and the boundary conditions are as follows

$$\varphi(t, 0, y) = 0, \varphi(t, 1, y) = 0, \varphi(t, x, 0) = 0, \varphi(t, x, 1) = 0 \tag{28}$$

$$\omega(t, 0, y) = 0, \omega(t, 1, y) = 0, \omega(t, x, 0) = 0, \omega(t, x, 1) = 0 \tag{29}$$

$$T(t, 0, y) = 0, T(t, 1, y) = 0, T(t, x, 0) = 0, T(t, x, 1) = 0 \tag{30}$$

and for hot zones [8–14]

$$\frac{\partial T}{\partial y} = 1 \tag{31}$$

2.4 Numerical method

To solve the system, the Method of Alternate Directions Implicit (ADI) was used [15]. Regarding the discretization of the equations, the finite difference scheme is used. For the time derivative that appears on the left-hand side of Eq. (7), we have

$$\frac{\partial T}{\partial t} = \frac{T_{i,j} - T_{i,j}^{n}}{\Delta t} \tag{32}$$

In Eq. (32), $T_{i,j}$ is the value of T at point (i, j) at the present instant, while $T_{i,j}^{n}$ is the value of T at point (i, j) at the previous time instant. For the case of the spatial derivatives found in the first term on the right-hand side of (Eq. (6)), we have

$$\frac{\partial T}{\partial x} = \frac{T_{i+1,j} - T_{i-1,j}}{2\Delta x} \tag{33}$$

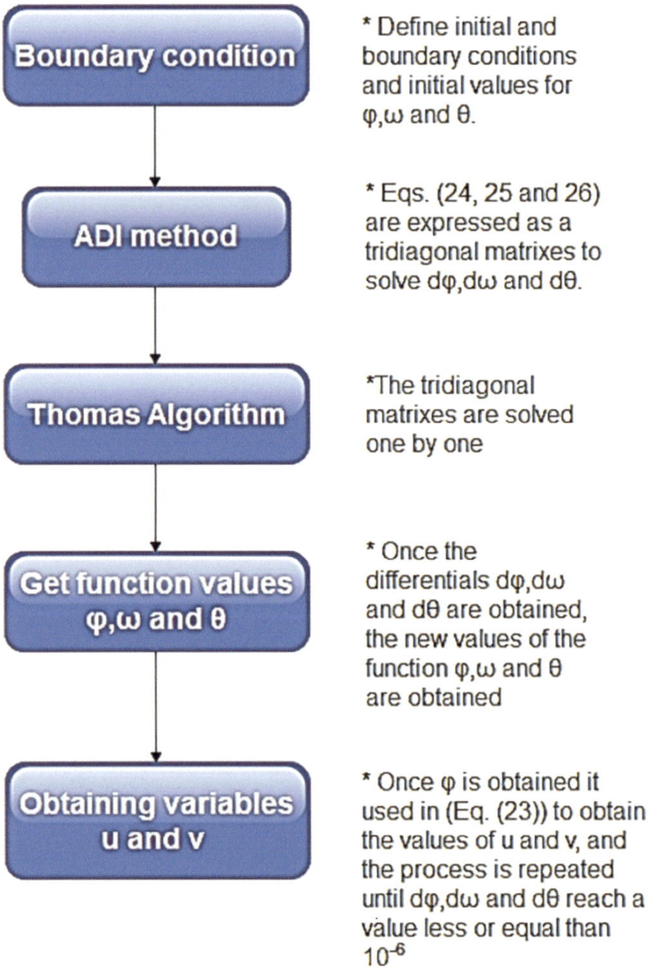

Figure 5.
General algorithm for the Method of Alternate Directions Implicit.

DOI: http://dx.doi.org/10.5772/intechopen.110295

Subroutine for φ :

```
!————————————————————————————————
SUBROUTINE funciondecorriente (y,dx,dy,dx2,dy2,phi,dphi,omega,u,v)
USE Constantes
IMPLICIT NONE
REAL (kind=8), DIMENSION(mi) :: dx,dx2
REAL (kind=8), DIMENSION(nj) :: y,dy,dy2
REAL (kind=8), DIMENSION(mi) :: ai,bi,ci,ri,ui
REAL (kind=8), DIMENSION(nj) :: aj,bj,cj,rj,uj
REAL (kind=8), DIMENSION(mi,nj) :: v,u,res,omega,chi
REAL (kind=8), DIMENSION(mi,nj) :: phi,dphi,phiant
DO i= 2,mi-1
DO j= 2,nj-1
u(i,j) = (phi(i,j+1)-phi(i,j-1))/2.d0*dy(j)
v(i,j) = -(phi(i+1,j)-phi(i-1,j))/2.d0*dx(i)
res(i,j)= dti*((phi(i+1,j)-2.d0*phi(i,j)+phi(i-1,j))/dx2(i) &
& + ((phi(i,j+1)-2.d0*phi(i,j)+phi(i,j-1))/dy2(j)) &
& + omega(i,j))
END DO
END DO
DO j= 1,nj
bi(1) = 1.d0
ci(1) = 0.d0
ri(1) = 0.d0 - phi(1,j)
ai(mi) = 0.d0
bi(mi) = 1.d0
ri(mi) = 0.d0 - phi(mi,j)
DO i= 2,mi-1
ai(i) = -dti/dx2(i)
bi(i) = 1.d0 + (2.d0*dti/dx2(i))
ci(i) = -dti/dx2(i)
ri(i) = res(i,j)
END DO
CALL tri(ai,bi,ci,ri,ui,mi)
DO i= 1,mi
chi(i,j)=ui(i)
END DO
END DO
DO i= 1,mi
bj(1) = 1.d0
cj(1) = 0.d0
rj(1) = 0.d0 - phi(i,1)
aj(nj) = 0.d0
bj(nj) = 1.d0
rj(nj) = 0.d0 - phi(i,nj)
DO j= 2,nj-1
aj(j) = -dti/dy2(j)
bj(j) = 1.d0 + (2.d0*dti/dy2(j))
cj(j) = -dti/dy2(j)
rj(j) = chi(i,j)
END DO
CALL tri(aj,bj,cj,rj,uj,nj)
DO j= 1,nj
dphi(i,j) = uj(j)
END DO
END DO
DO i= 1,mi
DO j= 1,nj
phi(i,j)= phi(i,j) + dphi(i,j)
END DO
END DO
END SUBROUTINE funciondecorriente
!————————————————————————————————
```

Table 1.
Subroutine for φ.

```
Subroutine for ω:
!
SUBROUTINE vorticidad (dx,dy,dx2,dy2,phi,u,v,theta,omega,domega)
USE Constantes
IMPLICIT NONE
REAL (kind=8), DIMENSION(mi) :: dx,dx2
REAL (kind=8), DIMENSION(nj) :: dy,dy2
REAL (kind=8), DIMENSION(mi) :: ai,bi,ci,ri,ui
REAL (kind=8), DIMENSION(nj) :: aj,bj,cj,rj,uj
REAL (kind=8), DIMENSION(mi,nj) :: v,u,res,phi,chi1,theta
REAL (kind=8), DIMENSION(mi,nj) :: omega,domega,omegant
DO i= 2,mi-1
DO j= 2,nj-1
res(i,j) = -(dti*u(i,j)*((omega (i+1,j)-omega (i+1,j))/(2.d0*dx(i)))) &
& -(dti*v(i,j)*((omega (i,j+1)-omega (i,j-1))/(2.d0*dy(j)))) &
& +((dti*Pr)*((omega(i+1,j)-2.d0*omega(i,j)+omega(i-1,j))/(dx2(i))))
&
& +((dti*Pr)*((omega(i,j+1)-2.d0*omega(i,j)+omega(i,j-1))/(dy2(j))))&
& -(dti*Ra*Pr*((theta(i+1,j)-theta(i-1,j))/(2.d0*dx(i))))
END DO
END DO
DO j= 1,nj
bi(1) = 1.d0
ci(1) = 0.d0
ri(1) = 2*(phi(1,j)-phi(2,j))/dx2(2) - omega(1,j)
ai(mi) = 0.d0
bi(mi) = 1.d0
ri(mi) = (2.d0*(phi(mi,j)-phi(mi-1,j))/dx2(mi)) -
omega(mi,j)
DO i= 2,mi-1
ai(i) = dti*(u(i,j)/(2.d0*dx(i))) - dti*Pr/(dx2(i))
bi(i) = 1.d0 + (dti*2.d0*Pr)/(dx2(i))
ci(i) = -dti*(u(i,j)/(2.d0*dx(i))) - dti*Pr/(dx2(i))
ri(i) = res(i,j)
END DO
CALL tri(ai,bi,ci,ri,ui,mi)
DO i= 1,mi
chi1(i,j)=ui(i)
END DO
END DO
DO i= 1,mi
bj(1) = 1.d0
cj(1) = 0.d0
rj(1) = 2.d0*(phi(i,1)-phi(i,2))/dy2(2) - omega(i,1)
aj(nj) = 0.d0
bj(nj) = 1.d0
rj(nj) = 0 - omega(i,nj)
DO j= 2,nj-1
aj(j) = dti*(v(i,j)/(2.d0*dy(j))) &
& -dti*Pr/(dy2(j))
bj(j) = 1.d0 + ((dti*2.d0*Pr)/(dy2
(j)))
cj(j) = -dti*(v(i,j)/(2.d0*dy(j))) &
& -dti*Pr/(dy2(j))
rj(j) = chi1(i,j)
END DO
CALL tri(aj,bj,cj,rj,uj,nj)
DO j= 1,nj
domega(i,j) = uj(j)
END DO
END DO
DO i= 1,mi
DO j= 1,nj
omega(i,j)= omega(i,j) + domega(i,j)
END DO
END DO
END SUBROUTINE vorticidad
!
```

Table 2.
Subroutine for ω.

```
Subroutine for T:
!———————————————————————
SUBROUTINE energia (dx,dy,dx2,dy2,u,v,theta,dtheta)
USE constantes
IMPLICIT NONE
REAL (kind=8), DIMENSION(mi) :: dx,dx2
REAL (kind=8), DIMENSION(nj) :: dy,dy2
REAL (kind=8), DIMENSION(mi) :: ai,bi,ci,ri,ui
REAL (kind=8), DIMENSION(nj) :: aj,bj,cj,rj,uj
REAL (kind=8), DIMENSION(mi,nj) :: v,u,res,chi2
REAL (kind=8), DIMENSION(mi,nj) :: theta,dtheta,thetant
DO i= 2,mi-1
DO j= 2,nj-1
res(i,j) = -dti*(u(i,j)*((theta(i+1,j)-theta(i-1,j))/(2.d0*dx(i)))) &
& -dti*(v(i,j)*((theta(i,j+1)-theta(i,j-1))/(2.d0*dy(j)))) &
& +(dti*(((theta(i+1,j)-2.d0*theta(i,j)+theta(i-1,j))/(dx2(i))) &
& +((theta(i,j+1)-2.d0*theta(i,j)+theta(i,j-1))/(dy2(j)))))
END DO
END DO
DO j= 1,nj
bi(1) = 1.d0
ci(1) = 0.d0
ri(1) = 0.d0 - theta(1,j)
ai(mi) = 0.d0
bi(mi) = 1.d0
ri(mi) = 0.d0 - theta(mi,j)
DO i= 2,mi-1
ai(i) = dti*(u(i,j)/(2.d0*dx(i)))-(dti/(dx2(i)))
bi(i) = 1+((dti*2.d0)/(dx2(i)))
ci(i) = -dti*(u(i,j)/(2.d0*dx(i)))-(dti/(dx2(i)))
ri(i) = res(i,j)
END DO
CALL tri(ai,bi,ci,ri,ui,mi)
DO i= 1,mi
chi2(i,j)=ui(i)
END DO
END DO
DO i= 1,mi
IF(configura==1) THEN
IF(i<(mi/3)) THEN
bj(1) = 1.d0
cj(1) = 0.d0
rj(1) = 0.d0 - theta(i,1)
ELSE IF(i>=(mi/3).and.i<=(4*mi/9)) THEN
bj(1) = 1.d0
cj(1) = -1.d0
rj(1) = 1.d0 - (theta(i,1)-theta(i,2))
ELSE IF (i>4*mi/9.and.i<(5*mi/9))THEN
bj(1) = 1.d0
cj(1) = 0.d0
rj(1) = 0.d0 - theta(i,1)
ELSE IF (i>=5*mi/9.and.i<=(2*mi/3))THEN
bj(1) = 1.d0
cj(1) = -1.d0
rj(1) = 1.d0 - (theta(i,1)-theta(i,2))
ELSE IF (i>2*mi/3)THEN
bj(1) = 1.d0
cj(1) = 0.d0
rj(1) = 0.d0 - theta(i,1)
END IF
END IF
IF(configura==2) THEN
IF(i<(mi/9)) THEN
bj(1) = 1.d0
cj(1) = 0.d0
rj(1) = 0.d0 - theta(i,1)
ELSE IF(i>=(mi/9).and.i<=(2*mi/9)) THEN
bj(1) = 1.d0
cj(1) = -1.d0
rj(1) = 1.d0 - (theta(i,1)-theta(i,2))
ELSE IF (i>2*mi/9.and.i<(3*mi/9))THEN
bj(1) = 1.d0
cj(1) = 0.d0
rj(1) = 0.d0 - theta(i,1)
ELSE IF (i>=mi/3.and.i<=(4*mi/9))THEN
bj(1) = 1.d0
cj(1) = -1.d0
rj(1) = 1.d0 - (theta(i,1)-theta(i,2))
ELSE IF (i>4*mi/9.and.i<(5*mi/9))THEN
bj(1) = 1.d0
cj(1) = 0.d0
rj(1) = 0.d0 - theta(i,1)
ELSE IF (i>=5*mi/9.and.i<=(2*mi/3))THEN
bj(1) = 1.d0
cj(1) = -1.d0
rj(1) = 1.d0 - (theta(i,1)-theta(i,2))
ELSE IF (i>2*mi/3.and.i<(7*mi/9))THEN
bj(1) = 1.d0
cj(1) = 0.d0
rj(1) = 0.d0 - theta(i,1)
ELSE IF (i>=7*mi/9.and.i<=(8*mi/9))THEN
bj(1) = 1.d0
cj(1) = -1.d0
rj(1) = 1.d0 - (theta(i,1)-theta(i,2))
ELSE IF (i>8*mi/9)THEN
bj(1) = 1.d0
cj(1) = 0.d0
rj(1) = 0.d0 - theta(i,1)
END IF
END IF
aj(nj) = 0.d0
bj(nj) = 1.d0
rj(nj) = 0.d0 - theta(i,nj)
DO j= 2,nj-1
aj(j) = dti*(v(i,j)/(2.d0*dy(j)))-(dti/(dy2(j)))
bj(j) = 1+((dti*2.d0)/(dy2(j)))
cj(j) = -dti*(v(i,j)/(2.d0*dy(j)))-(dti/(dy2(j)))
rj(j) = chi2(i,j)
END DO
CALL tri(aj,bj,cj,rj,uj,nj)
DO j= 1,nj
dtheta(i,j) = uj(j)
END DO
END DO
DO i= 1,mi
DO j= 1,nj
theta(i,j)= theta(i,j) + dtheta(i,j)
END DO
END DO
END SUBROUTINE energia
!———————————————————————
```

Table 3.
Subroutine for T.

$$\frac{\partial T}{\partial y} = \frac{T_{i,j+1} - T_{i,j-1}}{2\Delta y} \tag{34}$$

$$\frac{\partial^2 T}{\partial x^2} = \frac{T_{i+1,j} - 2T_{i,j} + T_{i-1,j}}{\Delta x^2} \tag{35}$$

$$\frac{\partial^2 T}{\partial y^2} = \frac{T_{i,j+1} - 2T_{i,j} + T_{i,j-1}}{\Delta y^2} \tag{36}$$

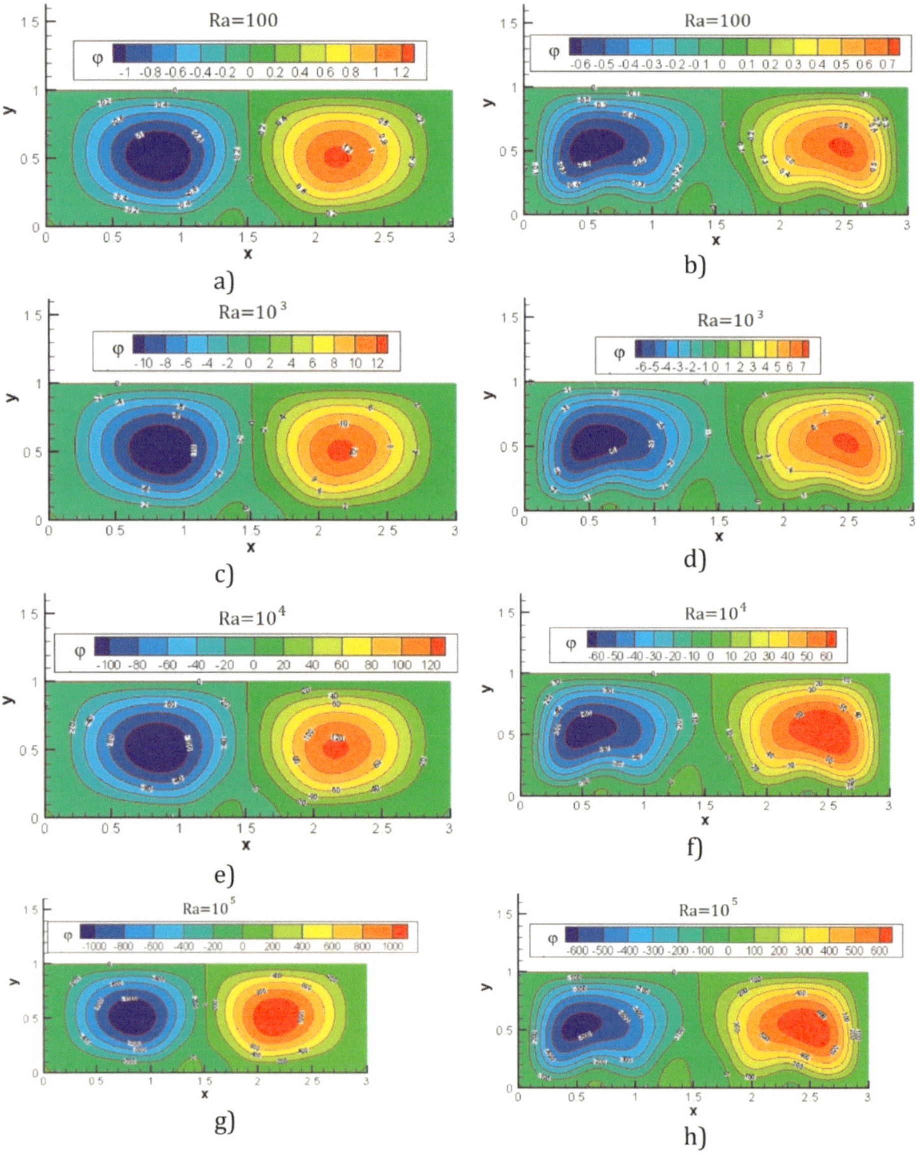

Figure 6.
Stream function φ for the proposed arrangements with different Rayleigh numbers. (a) Two heat sources and Ra = 1 × 10², (b) four heat sources and Ra = 1 × 10², (c) two heat sources and Ra = 1 × 10³, (d) four heat sources and Ra = 1 × 10³, (e) two heat sources and Ra = 1 × 10⁴, (f) four heat sources and Ra = 1 × 10⁴, (g) two heat sources and Ra = 1 × 10⁵, (h) four heat sources and Ra = 1 × 10⁵.

The numerical model was solved using a 200 × 100 uniform mesh and a time step $\Delta t = 1 \times 10^{-5}$. As a convergence criterion, it was established that $\Delta\varphi, \Delta\omega, \Delta T \leq 1 \times 10^{-6}$. The code was developed in the FORTRAN 90 programming language. It was compiled and executed using the commercial package Microsoft Visual Studio, using a 64-bit HP Pavilion 15-cw1xxx laptop, with an AMD Ryzen 53,500 U processor with a Radeon Vega video card. Mobile Gfx. [16–19].

The general algorithm is expressed as follows (**Figure 5**):

The subroutines used to solve Eqs. (28)-(30) are given in **Tables 1–3**.

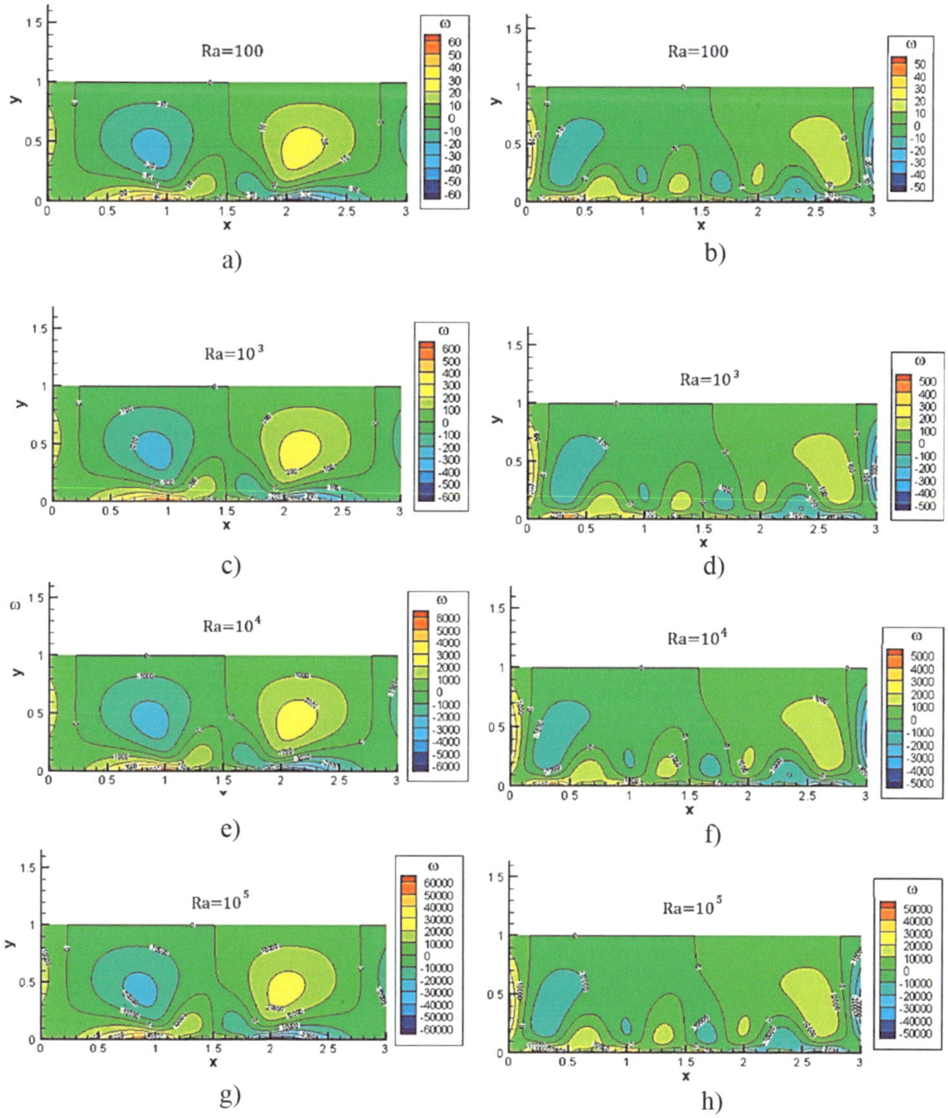

Figure 7.
Vorticity ω for the proposed arrangements with different Rayleigh numbers. (a) Two heat sources and Ra = 1×10^2, (b) four heat sources and Ra = 1×10^2, (c) two heat sources and Ra = 1×10^3, (d) four heat sources and Ra = 1×10^3, (e) two heat sources and Ra = 1×10^4, (f) four heat sources and Ra = 1×10^4, (g) two heat sources and Ra = 1×10^5, (h) four heat sources and Ra = 1×10^5.

3. Simulations and results

Before carrying out the simulations, the code used was validated using the fields of temperature and stream function obtained by Davis G., 1983.

As shown in the **Figures 6–8** corresponding to the profiles for both φ, ω, and T obtained, they replicate the results reported in the literature.

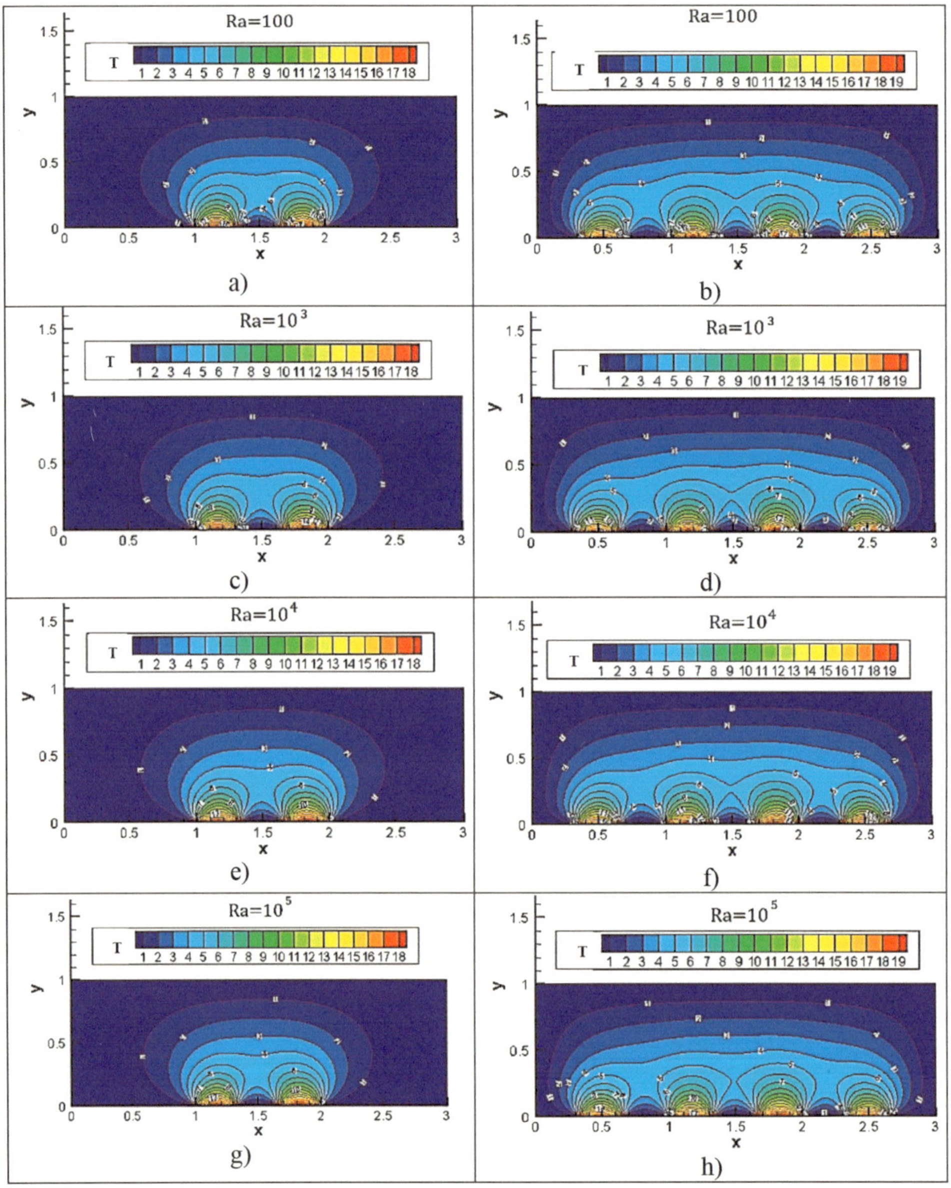

Figure 8.
Temperature T for the proposed arrangements with different Rayleigh numbers. (a) Two heat sources and Ra = 1×10^2, (b) four heat sources and Ra = 1×10^2, (c) two heat sources and Ra = 1×10^3, (d) four heat sources and Ra = 1×10^3, (e) two heat sources and Ra = 1×10^4, (f) four heat sources and Ra = 1×10^4, (g) two heat sources and Ra = 1×10^5, (h) four heat sources and Ra = 1×10^5.

All the simulations carried out were obtained with a fixed value of Pr = 7, with the characterization of the fluid as water. The computational experiments were performed with the two heat source configurations, using four values of the Ra number Ra (1×10^2, 1×10^3, 1×10^4, 1×10^5) obtaining the temperature distributions, as well as the formation of vorticity and the changes in the current lines.

Figures 6–8 show that, by keeping the fluid constant, with the greater number of Ra there is a greater T_H, i.e., the magnitude of q increases. The results obtained were found by establishing the following values for the constants $H = 0.2$ m, $\beta = 207 \times 10^{-6}$ K^{-1}, $k = 0.58$ W/mK $\alpha = 1.3882 \times 10^{-4} m^2/s$, $\nu = 1.004 \times 10^{-6}$ m^2/s, $T_0 = 25°C$.

Although the temperature profile does not express large changes for different values of Ra, the magnitude of the temperatures present in the system do.

The results of the simulations to reach the steady state times in the water heating process in the shrink tank show similar trends with the two arrangements of heat sources and very similar time values. For both arrangements, as the Rayleigh number increases, the times increase, but not significantly, having dimensionless time values that oscillate between 0.8647 and 0.9101 for the two and four arrangements respectively for the case of the Rayleigh value of 1×10^2. Y of 1.5503 and 1.6171 for the two and four arrangements respectively for Rayleigh = 1×10^5, as can be seen in **Figure 9**.

Regarding the temperature of the surface of the heat source, in the same way, similar trends are observed with the two arrangements of heat sources and very similar temperature values. But for Rayleigh values above 1×10^4, a difference is observed between both arrangements as can be seen in **Figure 10**. Obtaining values that go from 25 to 26°C for Rayleigh numbers = 1×10^2 and 1×10^3 respectively for both arrangements. However, for values of Rayleigh = 1×10^5 there are temperatures of 176.95 and 193.84°C for arrangements of two and four resistors respectively.

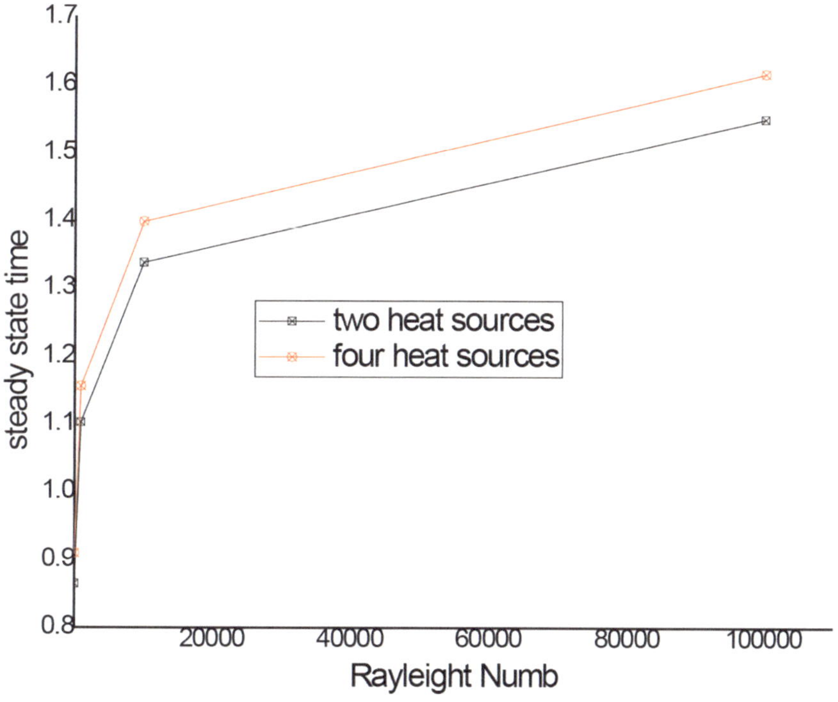

Figure 9.
Steady state time for the Rayleigh numbers of 1×10^2, 1×10^3, 1×10^4, and 1×10^5 for the two heat sources.

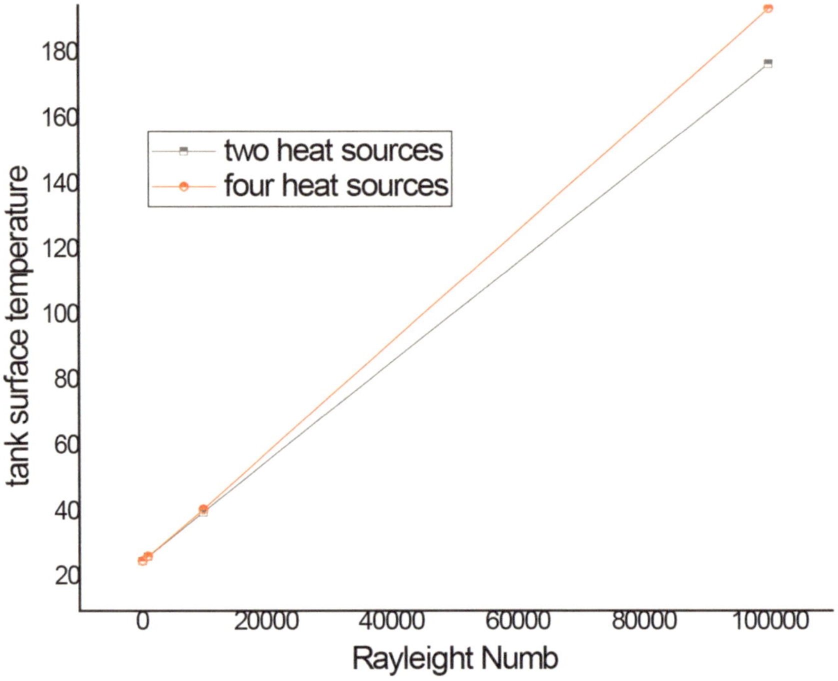

Figure 10.
Tank surface temperature for the Rayleigh numbers of 1×10^2, 1×10^3, 1×10^4, and 1×10^5 for the two heat sources.

In the same way, average temperature values of the water inside the shrink tank were obtained. There are similar trends with the two arrangements of heat sources and very similar temperature values below Rayleigh numbers = 1×10^3. But for values of Rayleigh $\geq 1 \times 10^5$ there are more marked temperature differences, as can be seen in **Figure 11**, for both arrays, but higher temperature values for the case of four arrays.

For the case of Ra = 1×10^2, which implies a heat source of q = 0.0024 W, there is a temperature near the heat sources of 25.1519°C on average, above ambient temperature. On the other hand, with a value of Ra = 1×10^4 with a heat source of q = 24.48 W, it produces a temperature of 176.9534°C, which is above the ambient temperature, however, in the areas closest to the heat sources this temperature can easily be exceeded.

From the above, we can say that, if we want an adequate equation through this study, it can be established that the heat sources, considering isothermal walls, must produce at least 25 W of heat, with the consideration that said sources, are in direct contact with the fluid.

On the other hand, the need to place four heat sources is contemplated, since, with the arrangement of two, it is insufficient to achieve a homogenization of the temperature in the heat shrink tank, generating areas where the temperature gradients are very small.

4. Conclusions

In the present work, computer simulations were carried out to evaluate the heat transfer in a shrink tank built in the facilities of the Faculty of Higher Studies Cuautitlán—UNAM.

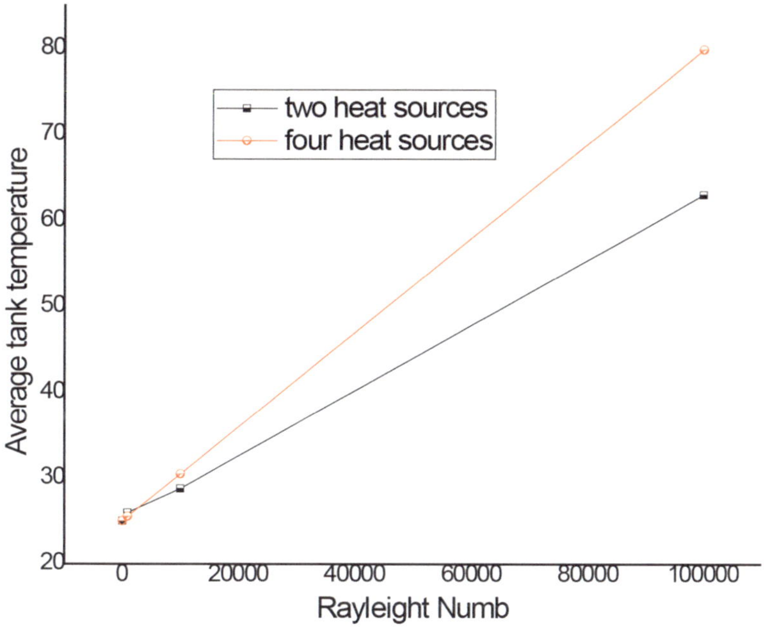

Figure 11.
Average tank temperature for the Rayleigh numbers of 1×10^2, 1×10^3, 1×10^4, and 1×10^5 for the two heat sources.

For the heating of the water in the tank, two different arrangements of electrical resistances were implemented as heat sources.

In order to quantify the heating times and temperature distributions, the Method of Alternate Directions Implicit was used in combination with the finite difference scheme, obtaining profiles of the stream function φ and vorticity ω, which helped with the selection of the resistance arrangement that guarantees a better heat transfer of the water in the tank.

From the simulations carried out in this work, it was observed that the formation of convective cells favors the homogenization of the temperature in the tank and that increasing the value of the Rayleigh number increases the vorticity but not the temperature field, which allows keep lower power.

It is confirmed that, when working with non-isothermal sources, a higher energy accumulation is obtained, but a less homogeneous temperature field than that produced by isothermal sources, deeper studies on this were carried out by Ostrach in 1988 [9].

The results of the simulations to reach the steady state times in the water heating process in the shrink tank, showed that the type of arrangement does not interfere directly in said time. Although it is observed that an increase in the Rayleigh number brings as a consequence an increase in the times to reach the steady state.

Likewise, regarding the temperature of the surface of the heat source, it can be seen that for values of the Rayleigh number above 1×10^3, the temperatures increase and there is a considerable difference with Ra = 1×10^5 for both arrangements, but higher for a four heat source arrangement.

Regarding the average temperature values of the water inside the shrink tank, for values of Rayleigh $\geq 1 \times 10^5$ there are more marked temperature differences between both arrangements, but higher temperature values are presented for the case of four arrangements, favoring the conditions required in the heat shrink process.

From the results obtained from the heating times with their respective temperature distributions, we can conclude that the arrangement that best optimizes the heat transport process in the shrink tank is the one corresponding to the four resistance arrangements, achieving a homogeneous temperature of 87°C, in times less than 9 min with a heat flux of q = 24.48 W.

The implementation of solid walls is considered for future work to study the effects of different insulation, in order to better conserve heat within the system, as well as the inclusion of the evaluation of heat transport in the shrink tub with the four arrangements, but incorporating a piece of meat with its shrink wrap, in order to quantify the distribution of temperatures with their respective validation with microbiological methods that indicate the null contamination of the product with bacteria.

Acknowledgements

The present work was developed under the sponsorship of the Facultad de Estudios Superiores Cuautitlán—Universidad Nacional Autónoma de México.

Conflict of interest

The authors declare no conflicts of interest regarding the publication of this paper.

Author details

José Luis Velázquez Ortega* and Aldo Gómez López
Facultad de Estudios Superiores Cuautitlán, UNAM, Estado de México, Mexico

*Address all correspondence to: siulj@unam.mx

References

[1] Potter NN, Hotchkiss JH. Food Science. 5nd ed. Springer; 1999. p. 623

[2] Velázquez OJL, López PJ, Cruz RP. Simulación con FEA de la transferencia de calor a través de un trozo de carne emulado, con geometría rectangular, en una cámara de termoencogido. Acta Universitaria. 2022;**32**:e3640. DOI: 10.15174. au.2022.3640

[3] Balaji C, Srinivasana B, Gedupudi S. Heat Transfer Engineering: Fundamentals and Techniques. London: Elsevier; 2021. p. 422

[4] Flynn AM, Akashige T, Theodore T. Kern's Process Heat Transfer. 2nd ed. USA: Wiley – Scrivener; 2019. p. 698

[5] Incropera F, De Witt D. Fundamentos de Transferencia de Calor. 4nd ed. México: Prentice Hall; 1999. p. 886

[6] Velázquez OJL. Principios de Transferencia de Cantidad de Movimiento. México: FESC – UNAM; 2018. p. 109

[7] Davis GD. Natural convection of air in a square cavity a bench mark numerical solution. International Journal for Numerical Methods in Fluids. 1983;**3**:249-264. DOI: 10.1002/fld.1650 030305

[8] AlAmiri A, Khanafer K. Bouyancy-induced flow and heat transfer in a partially divided square enclosure. Journal of Heat Transfer. 2009;**52**: 3818-3828. DOI: 10.1016/j. ijheatmasstransfer.2009.01.043

[9] Ostrach S. Natural convection in enclosures. Journal of Heat Transfer. 1988;**110**(4b):1175-1190. DOI: 10.1115/ 1.3250619

[10] Hartnett JKM, Kostic M. Heat transfer to Newtonian and non-Newtonian fluids in rectangular ducts. Advances in Heat Transfer. 1989;**19**: 247-356. DOI: 10.1016/S0065-2717(08) 70214-4

[11] Costa VAF. Thermodynamics of natural convection in enclosures with viscous dissipation. International Journal of Heat Mass Transfer. 2005;**48**(11): 2333-2341. DOI: 10.1016/j. ijheatmasstransfer.2005.01.004

[12] Pastrana D, Cajas JC, Treviño C. Natural convection and entropy generation in a large aspect ratio cavity with walls of finite thickness. In: Klapp J, Medina A, Cros A, Vargas CA, editors. Fluid Dynamics in Physics, Engineering and Environmental Applications. Springer, Berlin: Heidelberg; 2013. pp. 309-320. DOI: 10.1007/978-3-642-27723-8_27

[13] Gómez LA, García RBE, Vargas ARO, Martínez SLA. Mixed convection in a rectangular enclosure with temperature- dependent viscosity and viscous dissipation. In: Klapp J, Ruiz CG, Medina OA, López VA, Di GSL, editors. Selected Topics of Computational and Experimental Fluid Mechanics. Cham: Springer; 2015. pp. 253-259. DOI: 10.1007/978-3-319-11487-3_15

[14] Mil MR, Vargas R, Escandón J, Pérez RI, Turcio M, Gómez LA, et al. Thermal effect on the bioconvection dynamics of Gravitactic microorganisms in a rectangular cavity. Fluids. 2022;7(3): 113. DOI: 10.3390/fluids7030113

[15] Anderson DJ. Computational Fluid Dynamics. UK: Mc Graw-Hill Education; 1995. p. 547

[16] Chapra S, Canale R. Numerical Methods for Engineers. 7nd ed. USA: Mc Graw Hill; 2009. p. 968

[17] Cengel Y, Ghajar A. Heat and Mass Transfer: Fundamentals and Applications. 6nd ed. India: Mc Graw Hill; 2019. p. 1056

[18] Nieves HA, Domínguez SF. Métodos Numéricos Aplicados a la Ingeniería. 2nd ed. México: CECSA; 2006. p. 602

[19] Iriarte BR. Métodos Numéricos. 2nd ed. México: Trillas; 2012. p. 269

Chapter 6

Nature as a Teacher for Abiota Self-Organization in Terms of Entropy Analysis

Masoumeh Bararzadeh Ledari and Reza Bararzadeh Ledari

Abstract

In this chapter, the various terms of entropy generation in terrestrial systems and the atmosphere are estimated by imitating the entropy analysis of a steam power generation (STPG). The highest entropy generation is associated with the outgoing longwave radiation flux (more than 20–200 times the downward solar radiation). The results indicate that the most significant terms of entropy generation (heat dissipation) in different processes are related to latent and sensible heat fluxes (similar to steam generation and flue gas of the STPG). The vegetation cover (boiler system) destroys a part of solar energy absorption in the form of entropy generated by the formation of water vapor and transpiration (steam turbine). Given that life is formed by the optimal balance between the system, the ecosystem, and the living and nonliving organisms, it is important to study the various entropy fluxes in ecosystems that can lead to ecosystem balance.

Keywords: entropy analysis, nature ecosystem, Negentropy of the ecosystem, thermodynamics of ecosystem, Carnot efficiency of the nature

1. Introduction

Climate change and its effects on human life have shown that the identification of the material and energy flows interactions of nature, and the analysis methods of the natural phenomena are essential to achieve some new way to increase the ecosystem's adaptive capacity [1]. On the other hand, nature is an awesome system for humans; the source from which the best mechanisms and engineering ideas can be extracted. There are two fundamentally different ways of enabling humans to draw free energy. Firstly, low-entropy food produced by farming and photosynthesis is used to meet the metabolic needs of billions of people; in fact, the total amount of energy released by human metabolism can be compared with the energy that drives oceanic circulation. In a second way, low-entropy sources of energy such as fossil fuels, etc. are used by humans to maintain their external activities such as manufacturing, heating, etc. Energy consumption in this domain can surpass 10–100 times that of human metabolism [2].

One of the first studies done on the energy balance on the Earth is the valuable research work of Hartmann et al. [3]. In his climate studies, he conducted an overview of the atmospheric radiation budget measurement and found that the net flow of radiant energy between the upper atmosphere and the earth's surface is balanced.

In nature, what is almost always paramount is efficiency; there is no pattern of stability in that from which at least a single thing cannot be learned. Over the years, natural systems have been observed flawlessly, and all structures in this system have been in a good place. The problem with this statement is that the 2nd Law of thermodynamics allows systems to achieve order through working and using free energy in different cycles [4]. Irreversible work changes the order of the system. Heat transfer has the same effects that depend on the temperature gradient, both Irreversible work and heat transfer increases the entropy of the system; In other words, the Maximization of entropy production hinders ecological succession. Thermal analysis of ecosystems has shown that more mature systems store more solar energy, and have less heat dissipation. So, in these systems, the surface temperature will be colder and the temperature variance will be less [5]. Moreover, under the same environmental conditions, highly self-organizing systems must reflect solar radiation at a lower exergy level. It shows a cooler surface temperature. The hypothesis that older ecosystems have lower surface temperatures has been studied by only a few studies, despite the obvious applications for environmental management in the context of climate change and global warming [6]. Some researchers have concluded that tropical seasonal rainforests have lower levels of self-organization than adjacent farms, and the higher the daily average, the greater the degree of energy absorption and dissipation [7, 8].

Schneider and Kay [9] suggested that the incoming solar radiation is degraded; besides, it increases entropy. This process leads to maximum entropy production during ecosystem evolutions. Steinborna and Svirezhev [10] supposed the "entropy pump" hypothesis, which is a metric that quantifies the anthropogenic activities in an ecosystem. They studied the increase in entropy resulting from agricultural production; then, they concluded that the entropy generated by agricultural overproduction leads to less sustainability in the ecosystem; actually, Vast interactions in nature increase the free energy and cause significant thermodynamic disequilibrium in the atmosphere [11].

Erwin Schrödinger (1887–1961) highlighted negative entropy as a capacity of producing order out of disorder [9]. Negentropy is a thermodynamic phenomenon that leads to higher efficiency; nevertheless, the absolute magnitude of general efficiency cannot be determined due to the countless biotic and abiotic interactions in the ecosystem [12]. The equilibrium between biotic and abiotic structures is evolving during the earth's life and it generates entropy increment in some structures while it makes orders in others (negative entropy) [13]; In other words, Negentropy might be characterized as the stored energy in a highly structured system in terms of space and time. Negentropy describes a harmonically coupled series of causes and effects, where the total sum of harmonic effects is more intensively coupled than the original causes [14]. In this regard, Norris et al. has studied the effects of biomass functional diversity and solar degradation enhancement in terms of thermodynamic efficiency [15].

The occurrence of various reactions might lead to the formation of highly ordered structures. During this process, low-entropy flows out of the Sun, after dissipating the heat and interacting with the earth, turn into a high-entropy flux [16, 17]. Irreversible processes in the atmosphere are defined as net entropy fluxes, and the greenhouse effects are not deniable on their entropy flux [18]. The growth rate of short-wave entropy flux is higher than the growth rate of long-wave radiation flux. Therefore, with the absorption of more short-wave radiations, the irreversibility is further decreased.

Despite the potential of the second law of thermodynamics for ecosystem evaluation, lower researchers have used this concept to present the variation in ecosystem quality. In this chapter, a method to imitate the physical energy system has been suggested to measure the ecosystem quality changes. we deduce the entropy flux of different flows of the ecosystems in different climatic zones changes with seasons. The ability of the ecosystem as a self-organization system has been studied based on the negentropy of some interactions. For further understanding, to estimate the entropy efficiency of the ecosystem, these systems have been converted to the Carnot engine and Steam powerplants systems.

2. Materials and methods

Entropy is a state property that indicates the level of disorder in the system; its change between states can be calculated by the integral ratio of the reversible heat transfer to the absolute temperature. This ‘state of disorder” is characterized by the amount of disordered energy and its temperature level. In the case of reversible heat transfer between two systems, both systems would be at the same temperature and the amount of increase in one’s disorder matches the amount of decrease in the other’s disorder [19].

Here, the difference between net radiation and ground heat flux (G) can be defined as the available energy that can be calculated by the summation between SH and LE; though, due to the imbalance of surface energy, the observed available energy would be more often larger than the sum of EC-measured SH and LE. The energy balance closure ratio (EBR) accounts for 60–90% in most instances [20].

Entropy generation within a system creates internal irreversibility; Hence, no matter how the changes in the entropy of the system and its surroundings might be, the total entropy change (entropy generation) cannot be less than zero for any process [16].

As shown in **Figure 1**, human activities are part of the process of entropy generation, and nature should be able to adapt to the changes that have taken place.

In this research, we have developed a method thereby measuring entropy production in the ecosystem. The general equation of entropy balance is based on different terms of ecosystem flows; Here, different entropy fluxes in the ecosystem have been taken into account.

a. System boundaries and energy balance equation [21]

$$K_n - L_n - H - LE - dG = 0 \tag{1}$$

b. Second law of the entropy equation [21]

$$\dot{S}_{irrev} = \frac{dS_{sys}}{dt} + \sum_{out} \dot{m}_e s_e - \sum_{in} \dot{m}_i s_i - \sum \frac{\dot{Q}_j}{T_j} \tag{2}$$

$$\dot{S}_{irrev} = \frac{dS_G}{dt} + \dot{S}_H + \dot{S}_{LE} + \dot{S}_{K_{out}} + \dot{S}_{L_{out}} - \dot{S}_{K_{in}} - \dot{S}_{L_{in}} \tag{3}$$

c. Entropy of radiation [21].

$$\dot{S}_{L_{in}} = \frac{L_{in}}{T_{sky}} \text{ and} \tag{4}$$

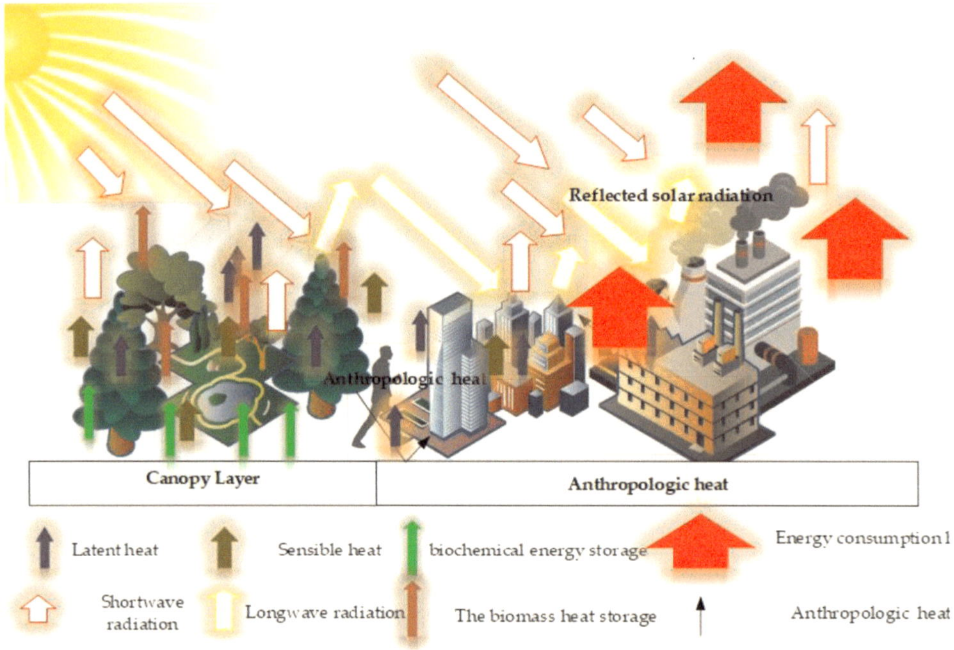

Figure 1.
The main entropy flows of nature.

$$\dot{S}_{K_{in}} = \frac{K_{in}}{T_{br}} \tag{5}$$

$$\dot{S}_{K_{in}} = \left(C_{d\ ir}\, p_{dir}^{0.9} + C_{diff}\left(1 - p_{dir}\right)^{0.9}\right)K_{in}^{0.9} \tag{6}$$

$$\dot{S}_{L_{out}} = \frac{L_{out}}{T_{surf}} \tag{7}$$

d. Entropy of sensible and latent heat flows [21]

$$\dot{S}_{H} = \frac{H}{T_{air}} \tag{8}$$

$$\dot{S}_{LEheat} = \frac{LE}{T_{air}} \tag{9}$$

$$\dot{S}_{LEmix} = (E)(R_v)\ln(RH_{amb}) \tag{10}$$

e. Storage terms and steady-state assumptions [21]

$$\frac{dS_{G_i}}{dt} = \frac{dG_i}{dt}\frac{1}{T_i} \tag{11}$$

f. the Carnot efficiency of the biosystem can be calculated as [22]:

$$\eta_{biosystem} = \frac{T_{Ave.air}}{T_{Sun}} \tag{12}$$

$$W = Q_H - Q_C \tag{13}$$

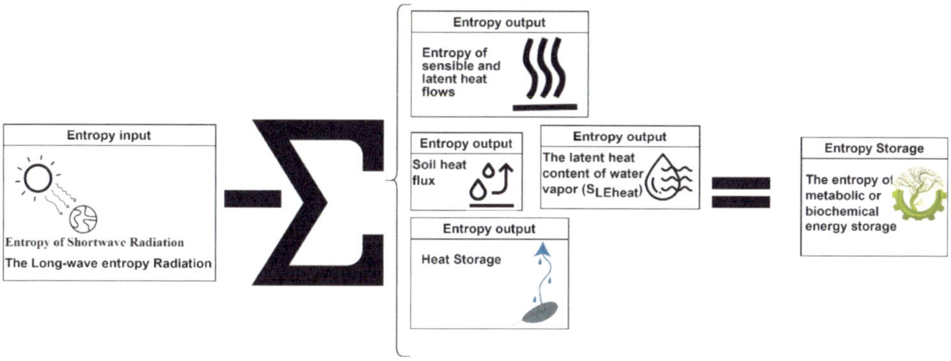

Figure 2.
Different interactions between the ecosystem's components in terms of the entropy balance of the ecosystem.

Regarding different interactions between ecosystem's components, **Figure 2** has been illustrated the entropy balance of the ecosystem.

3. Results and discussions

Sun radiation contains short-wave and long-wave radiation. Short-wave radiation has a considerable amount of energy; whereas, long-wave radiation has a lower energy content. UV rays from short-wave radiation passing through the atmosphere are absorbed by the clouds and the earth's surface. Part of the energy received by the surface is reflected into the atmosphere; this energy is re-emitted to the atmosphere as infrared rays. Here, the entropy balance between different flows that come and go to/from the earth has been studied (**Figure 1**).

The change of seasons in different environments, depending on the latitude and geometric orbit of the Earth, creates changes in solar radiation. Due to the diversity of climates and different land uses in Iran, four provinces have been considered in different climates. **Figure 3** shows the four studied areas and their land uses severally.

3.1 Entropy of shortwave radiation in different climatic zones

The energy of short-wave radiation is absorbed by the surface according to the geographical location (**Figure 4a–d**). Sunny days are more frequent in the center of Iran than in the north of the country. In the south of Iran, due to dust and humidity effects, sunny days are becoming less frequent. In the north, the cloudy days are so frequent that the short-wave radiation in this area is less than in the other areas. Generally, shortwave radiation is the main source of energy on the earth and can be changed based on the quality of the atmosphere and earth.

Cold radiation—the radiation of long-wave rays—is the main process of heat loss in the Earth's climate system. The balance between this energy loss (output) and the thermal energy resulting from short-wave radiation (input) determines global warming or global cooling of the earth's system.

In Mazandaran province, in winter, in comparison to other seasons, the entropy generation of surface downward shortwave radiation flow is about 10 times (**Figure 5a–d**). The earth's surface temperature is very low in winter, which means that much of the sun's radiation is reflected due to the lack of vegetation cover; thus,

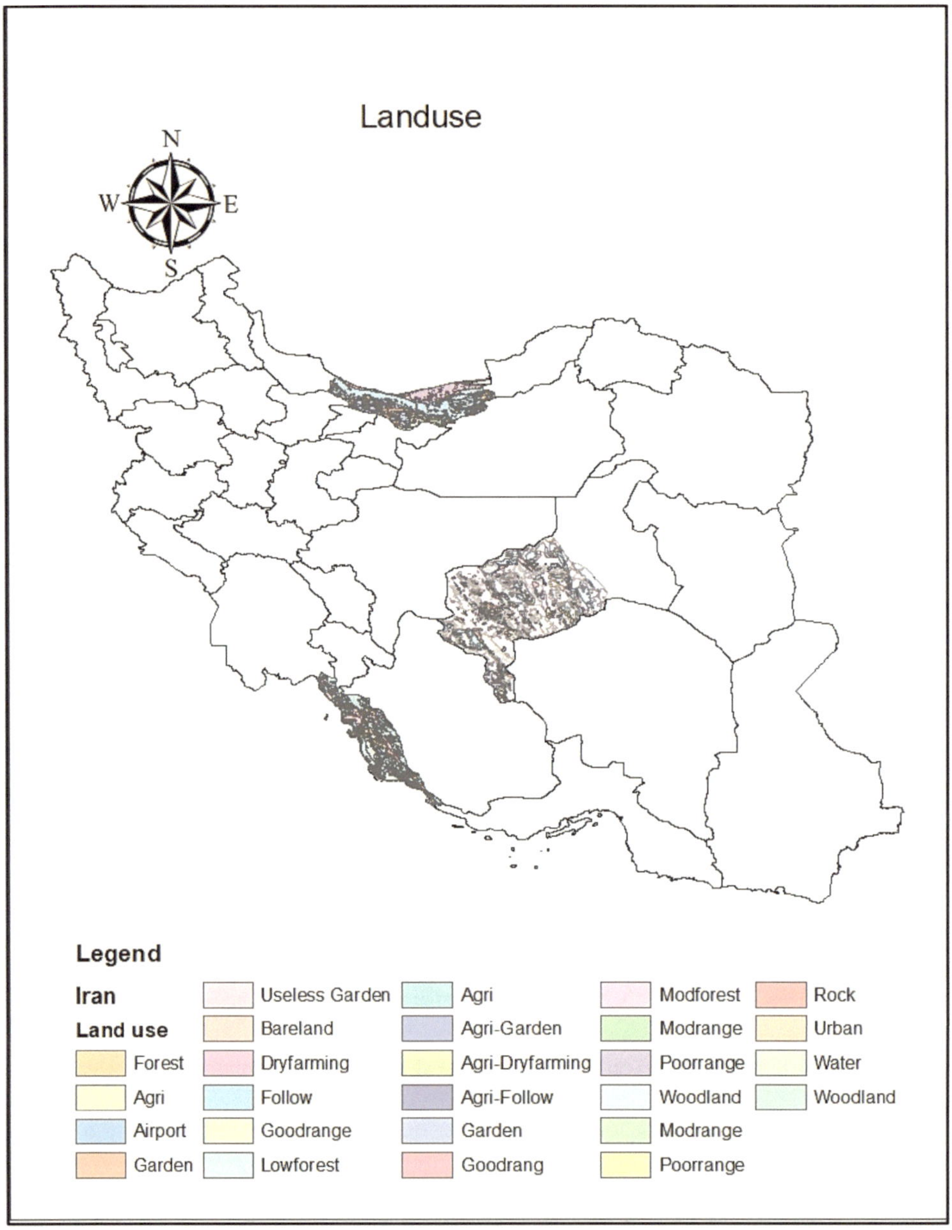

Figure 3.
Land use of different case studies in this study.

entropy production increases in this season. In other seasons, this value increases to about 12 W/m^2.

In Bushehr province, in spring, entropy production increases from 5.8–7.4 W/m^2. Compared to other seasons, in summer, radiation absorption does not have much effect on entropy production and it varies around 5.5–8 W/m^2. This is due to the high temperature and the ability of the earth to absorb the high levels of radiation owing to the position of the sun in this season. In autumn, the quality of the earth for sunlight

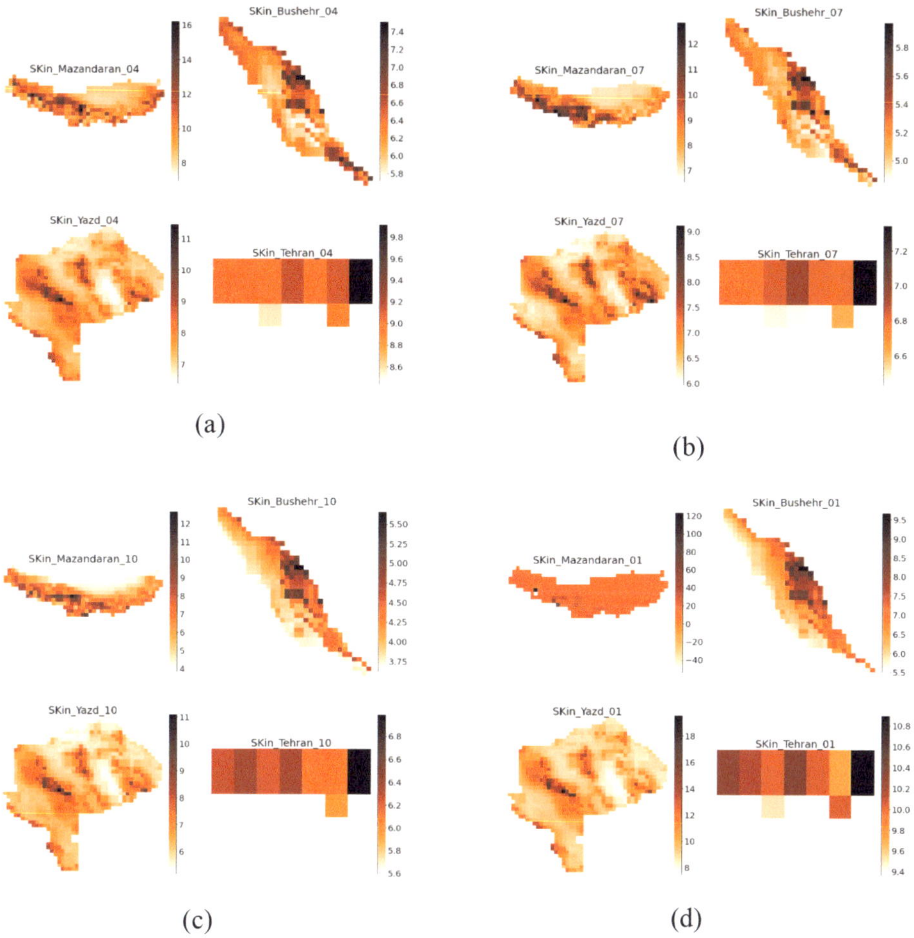

Figure 4.
The entropy of surface downward shortwave radiation flows in different climates in spring (a), summer (b), autumn (c), and winter (d) (W/m²).

absorption changes, and the absorption amount decreases from about 3.75 to 5.5 W/m^2. Interestingly, in winter, as a result of changes in temperature, entropy production on Earth increases. It seems that the Earth is getting prepared for the increase of its power in the winter season.

Yazd province is no exception to this rule. In winter, in Yazd, the entropy generation is about 7 W/m^2 more than anywhere else in autumn; however, this increase, compared to summer, is about 2 W/m^2 in fall. In spring, land developments behave differently; therefore, entropy production in this season is approximately 2 W/m^2 higher than in summer.

In Tehran province, the amount of entropy production in spring is about 8–10 12 W/m^2. This amount decreases to some extent in summer and reaches approximately 6.6 to 7.2 12 W/m^2. In autumn, Land cover and air temperature changes reduce this value to 5.6–7 12 W/m^2. In winter, the decrease in temperature intensifies the effect on entropy production; therefore, entropy production increases to about 9.4–10.8 12 W/m^2.

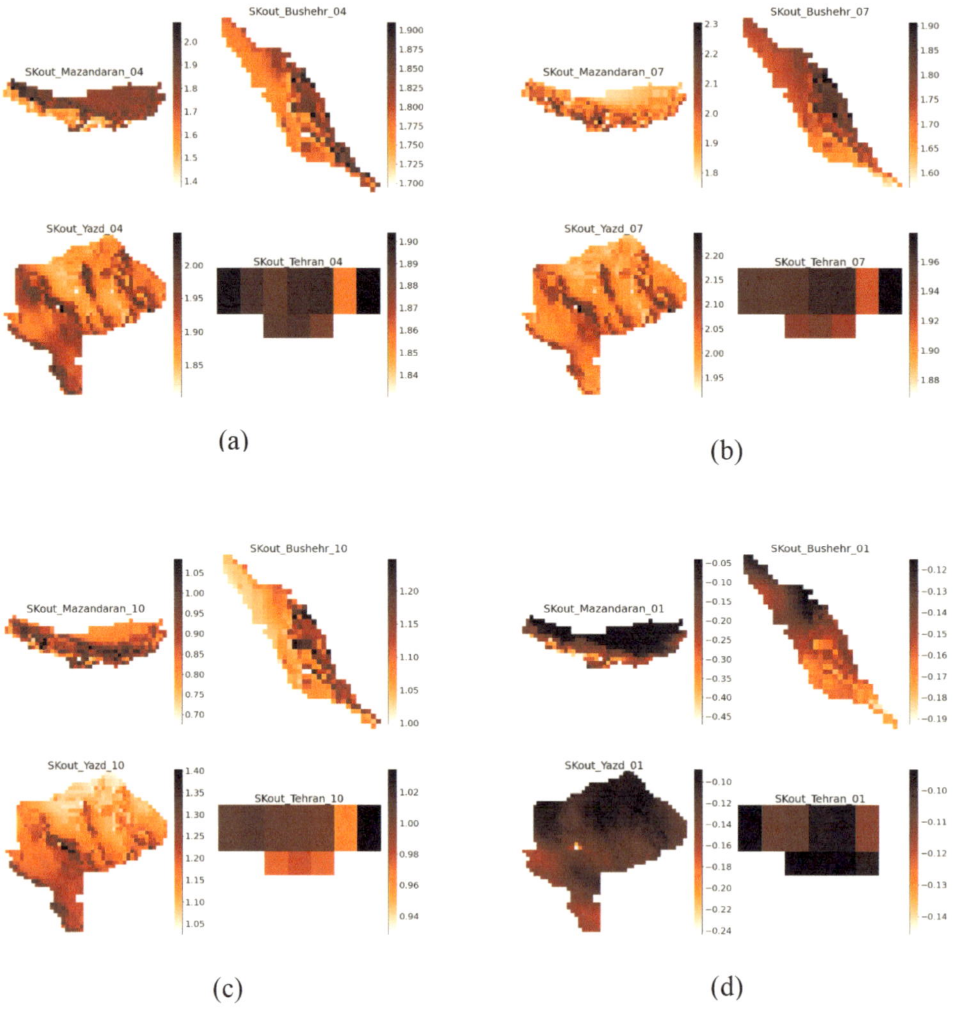

Figure 5.
The entropy of surface outgoing shortwave radiation flows in different climates spring (a), summer (b), autumn (c), and winter (d) (W/m²).

As **Figure 5d** shows in winter nature has negentropy. Negentropy can preserve the efficiency of the ecosystem. This amplifies the order and improves the general efficiency of the ecosystem.

In winter, the length of days is shorter than at night; that is, the surface outgoing shortwave radiation during nights increases and it leads to negentropy. It seems that radiative cooling which leads to the absorption of the long-wave radiation energy by the earth (infrared), can balance the short-wave radiation (visible light). In particular, both heat transfer convection and latent heat evaporation transfer are important in removing heat energy from the surface and redistributing that in the atmosphere. It is worth noting that daily diversity and geographical differences complicate the outgoing and downward radiation energy either.

In other seasons, the entropy of the surface outgoing short-wave radiation flows generates about 1–3 W/m^2 in each location.

As shown in the figures above, in general, the amount of shortwave radiation in areas with dense vegetation (a part of the studied provinces according to the land use map especially Mazandaran, and Yazd province) is more, while the temperature is lower. So, as can be seen in this sector's figures, these areas have more shortwave entropy radiation, and the opposite of this is seen for areas with sparse vegetation. The reason is that the dense vegetation cover areas have low surface albedo, and high evapotranspiration values, and vice versa. Moreover, the dense vegetation cover areas have higher latitudes, a decrease in the sun's angle, and the amount of cloudiness are also other reasons for this issue.

Also, winter albedos of treeless areas are higher than forested areas, because snow does not easily cover trees. The summer albedo is related to the amount of photosynthesis process because plants with high growth capacity have a larger fraction of their foliage to receive light. The result is that wavelengths radiations that are not used in photosynthesis are mostly reflected into space instead of being absorbed by other surfaces.

On the other hand, considering that the highest surface albedo is in the winter season and the minimum amount of reflection is in the spring season, therefore, the entropy caused by the absorbed radiation in the spring season is higher in all areas than in the winter season. Of course, in desert areas such as parts of Yazd and Tehran provinces, the amount of absorbed radiation is lower and therefore they produce less entropy of shortwave radiation, which is because deserts albedo is almost high.

3.2 The long-wave entropy radiation in different seasons and climates

The important climatic process of atmospheric warming takes place mainly through the earth's surface, which is heated during the absorption of solar energy and thus is itself a source of radiation. The altitude of different climatic zones is presented in **Table 1**.

Longwave entropy flux is determined by dividing the upper atmosphere solar irradiation from layers in the atmosphere to sky temperature. In this chapter, using NOAA satellite observation data sets, we study the entropy budget of the ecosystem during different months and seasons.

The amounts of infrared fluxes depend on different factors such as the thickness of the cloud (the thicker the cloud, the less heat escapes into space), the height of the cloud, the water vapor content of the atmosphere (less heat is released into the atmosphere from a highly wet atmosphere), water temperature and snow cover.

The northern areas of Mazandaran province are close to the sea and, compared with the southern areas that are covered by forests, they have fewer vegetation areas; therefore, the amount of radiation energy that returns to the atmosphere is higher in the northern areas, and consequently, more entropy flux is produced. Usually in

Province	Longitude	Latitude	Altitude
Mazandaran	50.9	36.7	36.73
Bushehr	50.8	28.9	28.95
Yazd	54.3	31.8	31.88
Tehran	51.3	35.7	35.7

Table 1.
Different case studies and their longitude, latitude, and altitudes.

summer, when the surface air is warmer, the amount of entropy production decreases to some extent. Proportionately, the entropy production process in autumn is somewhat similar to the process in summer. Due to the little and far vegetation in winter, most of the radiation flux is reflected from the ground. Another point is that nights in winter are longer than days. So, the entropy produced at night would have a greater effect on the total entropy flux in some areas. Since the earth is warmer in the day than at night and the direction of heat movement is in the opposite direction of the day, it is understood that in some parts of the earth, the amount of negative entropy generation at night is more than the amount of that at night. It occurs especially in winter.

The increase in the flux of downward long wavelength's entropy (due to the clouds and particles in the air) is almost several times greater than the wavelength radiation emitted from the atmosphere. In Mazandaran province, this amount increases to a great extent in some areas. In spring, due to the presence of clouds, this amount has a high value and affects humidity in Mazandaran. That is, in many areas, this value changes between 10 and 50 W/m^2. Only a limited part of the studied region has 3000 W/m^2; whereas, in summer, due to the decrease in cloudiness conditions, the amount of solar energy in the atmosphere decreases and changes by about 10–18 W/m^2. In fall, although in some limited areas this amount has reached about 10,000 W/m^2, the average value is between 0 and 2000 W/m^2. In winter, significant changes in this value are observed. and in many areas, the amount varies between 0 and 20 W/m^2 (**Figure 6a–d**).

In Bushehr region, the long-wave entropy generation in spring, from the energy input to the atmosphere is more than in other seasons, while these amounts decrease in winter and autumn.

In Bushehr province (**Figure 7a–d**), the rate of increase in incoming entropy to the atmosphere, in some seasons, is about 10–20 times higher than this value in Mazandaran province. In this region, dust generated from industrial activities, and the effects of air humidity, are the main items varying the entropy generation between 0 and 100 W/m^2. In spring and summer, this value is between 0 and 50 W/m^2 in many areas; however, in urban and industrial areas this amount increases to about

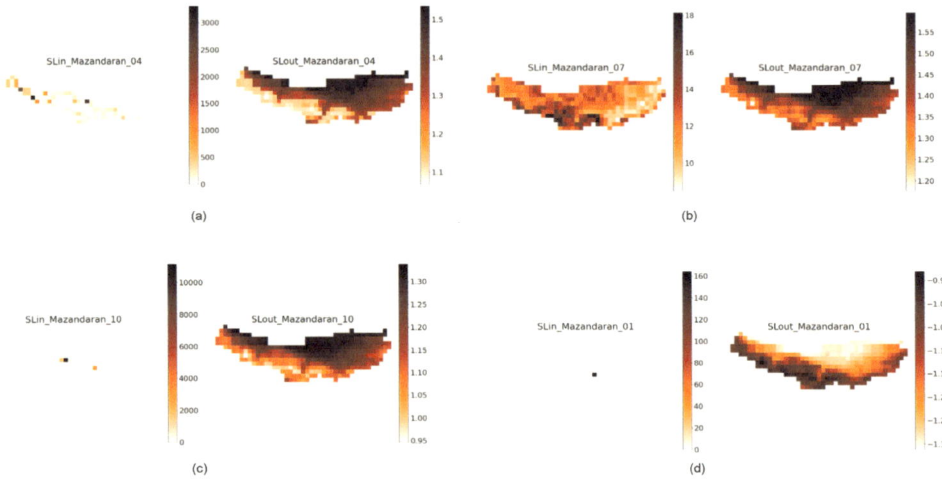

Figure 6.
The entropy content of the incoming and outgoing long-wave radiation fluxes (S_ Lin, S_ Lout) in a wet and cold climate (a), a wet and hot climate (b), a dry and hot climate (c), and a mountain climate (d).

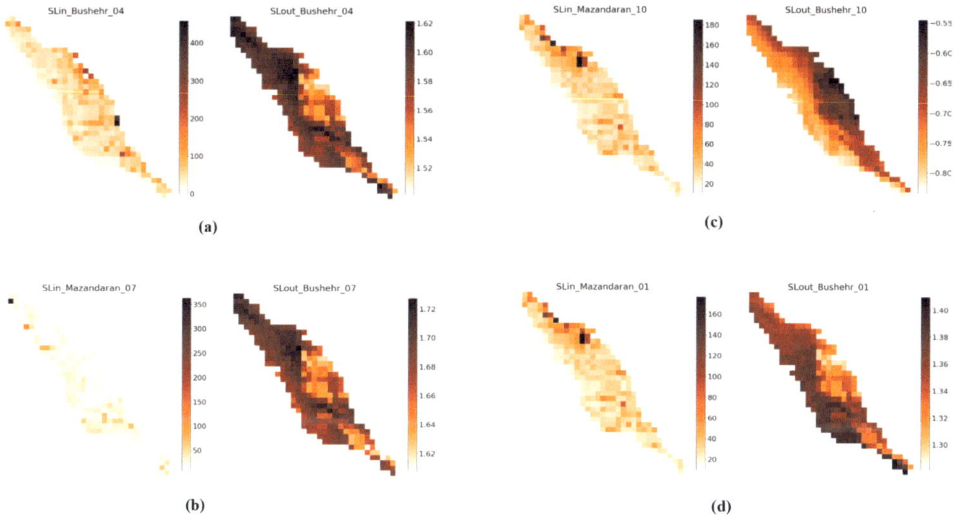

Figure 7.
The entropy content of the incoming and outgoing long-wave radiation fluxes (S_ Lin, S_ Lout) in a wet and cold climate (a), a wet and hot climate (b), a dry and hot climate (c), and a mountain climate (d).

400 W/m^2. In autumn the values of the entropy increase by about 80 W/m^2 though; that is, in summer the disorder created by the atmosphere is up to about 2 times less than that in spring. In winter, the previous trend seems to continue.

A large area of Yazd province is semi-desert and without vegetation cover. Moreover, due to the semi-mountainous topography, it has many slopes facing southwest, which increases the amount of radiation per unit area, and consequently, the amount of reflection [23].

Increased tourism effects in spring generated radiation entropy because of the effect of cloudiness and pollution (**Figure 8a**). In summer, these effects are reduced to about one-tenth of their value in spring, which may be due to a rising sky

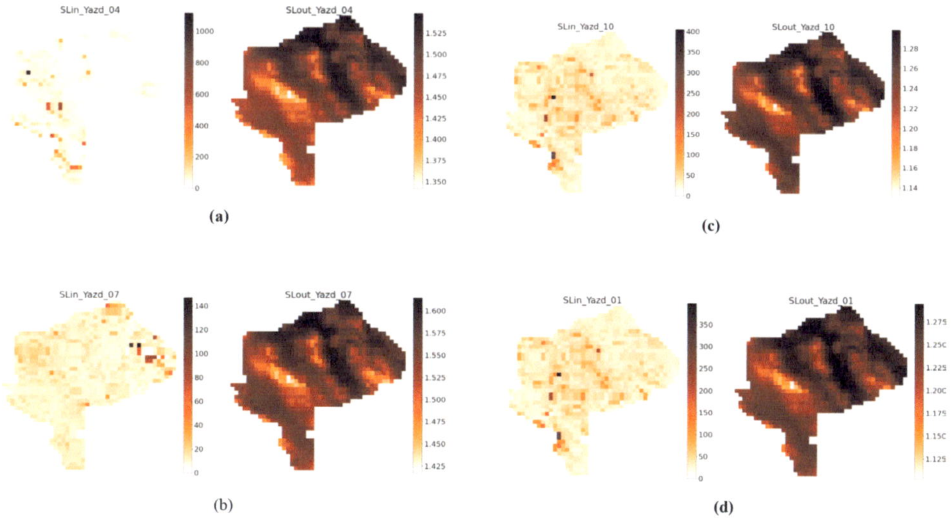

Figure 8.
The entropy content of the incoming and outgoing long-wave radiation fluxes (S_ Lin, S_ Lout) in a wet and cold climate (a), a wet and hot climate (b), a dry and hot climate (c), and a mountain climate (d).

temperature (**Figure 8b**). In autumn and winter, the trend of entropy increase is almost the same. As shown in **Figure 8a–d** urban areas have a much greater value compared to other areas, which can well show the effect of air pollution on solar radiation.

In spring, the amount of long-wavelength entropy production in the atmosphere varies in the range of 0–200 W/m^2 in most areas of Yazd province. Only a limited percentage of this region produces entropy up to about 600–1000 W/m^2.

In summer, some areas of Yazd province have the same value as in spring. Only in limited areas, where the average entropy production is about 600 W/m^2 in spring, this amount is reduced to about 100–140 W/m^2. In fall, the produced entropy is about 3.5 times that of summer. Finally, in winter, its amount does not change much compared to autumn and is almost constant.

In Tehran province, the effect of increased entropy due to the input long wavelength is very significant in spring. One of the important reasons for entropy production at the atmospheric level is the existence of pollutants in the atmosphere. Tehran, as a region with a high population density, is one of the most polluted cities in Iran. Therefore, entropy production in some areas of Tehran is worth considering, especially in the eastern part of Tehran, where there are a lot of car travel and industries. On the other hand, the degree of cloudiness in spring is much higher than that in summer, so the effect will be much greater. As shown in **Figure 9c** and **d**, similar behavior has been seen in autumn and winter.

In Tehran, the situation is completely different in different areas. In spring (**Figure 3(a)**), the generated entropy in western areas reaches about 1400 W/m^2 and in other areas, it is about 600 W/m^2 on average. In summer (**Figure 9b**), the amount of entropy production reaches an average of about 15 W/m^2. This value has increased 10 times in autumn and even in some areas- southeast of Tehran- it reaches 120 W/m^2. In winter, the same trend is observed as in autumn.

The up-wave radiation leads to an increase in the surface temperature of the earth. In Mazandaran, vegetation cover leads to Sunlight absorption, which has loosed energy from the plants and temperature change.

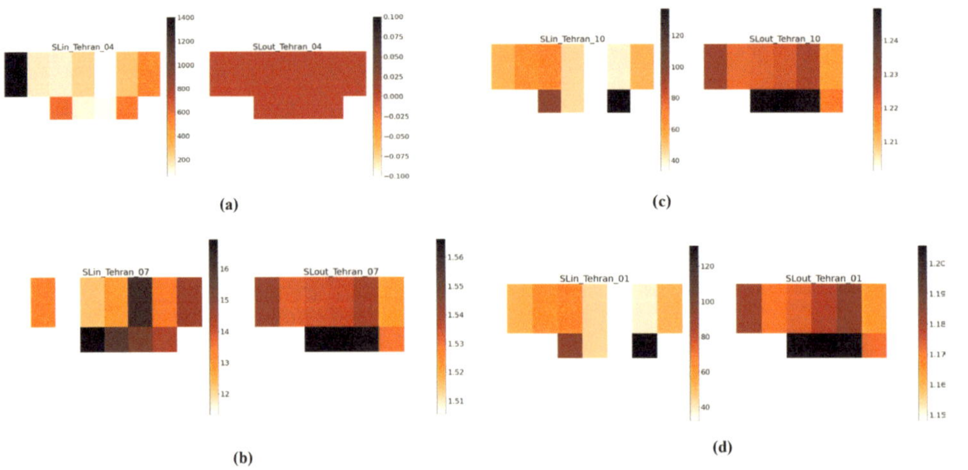

Figure 9.
The entropy content of the incoming and outgoing long-wave radiation fluxes (S_ Lin, S_ Lout) in a wet and cold climate (a), a wet and hot climate (b), a dry and hot climate (c), and a mountain climate (d).

Moreover, Entropy from long-wavelength radiation fluctuates between spring and summer seasons, as well as latitude variation. The long-wavelength entropy radiation is higher in the spring season and the southern regions of the country (Bushehr province), and it can be said that one of the reasons for the maximum entropy in this area is related to the maximum long-wave radiation from the south of Iran (Bushehr province). In the spring season, the vertical angle of the sun and the sky is clear, the amount of received energy is more than in the regions with higher latitudes, and accordingly, the amount of energy output is more in the south of Iran (Bushehr province) than in the north of Iran (Mazandaran province). In the southern regions of the country (Bushehr province), there has been an increase in the relative humidity in the atmosphere, and this humidity, like a greenhouse gas, plays as a hurdle for existing long wavelength radiations. In the summer season, the role of lower latitude and the effects of the country's mountains (Yazd province) increases, and as a result, the energy input and output from the earth's surface increases, the effects of which are evident in the increase of entropy generation in Yazd province in the summer season. One of the reasons for the output radiation of long wavelength in the south of Iran (Bushehr province), and southeast of Iran (Yazd province), in the winter season, is the vertical angle of the sun and the longer duration of the sun's radiation, which causes a lot of input energy to the earth's surface and the lack of clouds in this area causes a large part of the solar energy to escape. In the winter season, the oblique angle of the sun, the shortness of the day, and the humidity and cloudiness of the atmosphere in higher latitudes (Mazandaran and Tehran provinces) prevent the entry of the incoming energy of the sun and as a result, reduce the output of the sun's long wave radiation, and therefore, the entropy of the long wave radiation in these areas is less than others.

3.3 The latent heat content of water vapor (S_{LEheat})

Latent heat is a huge heat transfer in the atmosphere. It is a phase change of water that absorbs surface heat. Sensible and latent heat flows are important forms of heat that flow through the earth's energy balance. Water evaporation moves up and condenses in the atmosphere. This heat is higher than other heat flows on the ground.

Water vapor absorbs most of the infrared waves. The amount of atmospheric water vapor in summer is higher than that in winter due to higher temperatures (absolute humidity). Therefore, the amount of water vapor (under the same conditions) in summer is more than that in winter.

It is clear that winter has the lowest entropy generation in the form of sensible and latent heat flows in different climatic zones, which is related to the type of plant that can grow in these climatic zones in winter. It is undeniable that wet and cold climates have the highest value among different climatic zones. Mazandaran in the wet and cold climate is covered by forest and agricultural land, which is the main source of latent and sensible heat fluxes. In Bushehr, the surface area has almost a similar behavior in spring and summer, while in autumn changes in nature occur, leading to an increase in the generation of entropy from water evaporation (**Figure 10c**). Generally, in autumn, the entropy generation of latent and sensible heat has a little increase in both hot climates (wet and hot climate and dry and hot climate) in comparison to summer entropy generation (**Figure 10a** and **b**). Entropy generation in a dry and hot climate has a higher value in the center of Yazd. This is the Taft region near Shirkooh Mountain which is the highest region in Yazd province. Due to its altitude and rainfall, agricultural activities are the dominant economic activity in this

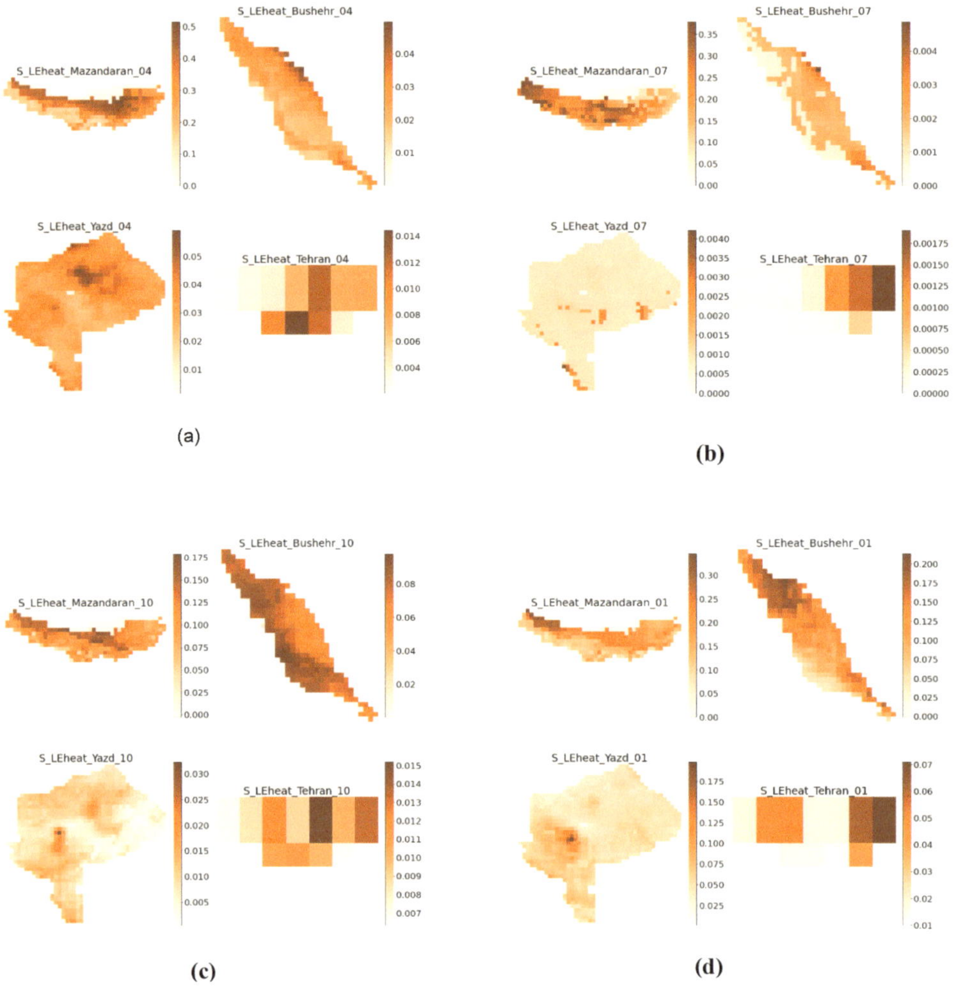

Figure 10.
The entropy of sensible and latent heat flows in different climates in spring (a), summer (b), autumn (c), and winter (d) (W/m^2).

region. So this area has a different behavior compared to other areas of this province. Entropy increment is more common in spring compared to other seasons in all climates. Winter decreases vegetation area, so, the ecosystem shows the least reaction in winter (**Figure 10d**).

Latent heat transport essentially couples the biosphere and atmosphere, as well as the mass and energy cycles associated with surface-atmosphere transport processes. Although vegetation growth by transpiration process attempts to establish local thermodynamic balance, this means that it maximizes the conductance of materials in the plant as well as greater productivity of the vegetation growth. As can be seen in the figures above, in areas with a dense vegetation cover, the amount of entropy caused by evapotranspiration is much higher than in other parts, and in the spring, in the northern regions of Iran (Mazandaran and Tehran provinces), and in the summer, South of the country, such as Bushehr and Yazd provinces, have the highest entropy due to high latent heat generation in line with increasing the vegetation growth. Latent heat transfer is also one of the primary processes related to entropy production in the atmosphere, and its amount is higher than other entropy productions in the atmosphere due to the phase change process.

3.4 Entropy of soil heat flux

Soil moisture is one of the most important soil variables that is widely used in the study and management of soil and water resources. It has a temporal and spatial nature and is one of the important components of climatic, ecological, and hydraulic models. Accordingly, the spatial distribution of soil moisture in different longitudes shows that the highest amount of soil moisture occurs at a longitude of 52°, and the lowest value occurs at longitudes of 46 and 62°. The maximum value also occurs at a latitude of 38°, and the lowest value occurs at latitudes of 30–32° [23].

To examine the amount of soil moisture, 133 images have been used, which were generated by the NASA-USDA Global Soil Moisture Data satellite in 2019 and early 2020.

As shown in **Figure 11**, Tehran has the lowest soil moisture content followed by Yazd and Bushehr. As can be seen, the soil moisture content increases everywhere in early spring. In summer, it decreases sharply and during autumn and winter, this value increases due to an increase in the amount of rainfall.

Soil heat flux is greatly influenced by net radiation at a certain depth of the ground [24]. In this regard, the NOAA satellite dataset has been utilized to prepare the required data (topsoil (0–10 cm)).

In some areas, the negative soil entropy is observable. This shows that a part of negative entropy in the earth is formed inside the soil. It means that the soil tends to order and there is the least amount of chaos. In some seasons, in a wet and cold climate, it is clear that the entropy increases due to the vegetation process happening in the soil (**Figure 12a** and **b**). In autumn, the earth receives more organic materials from things, such as fallen leaves, which cover the soil surface (**Figure 12c**). Since the microorganism process is so slow, the effects of this reaction during a month are not considerable. So, the soil system can move towards sustainability. The wet and hot climate generates more entropy. This area is mostly covered with soil and does not have any conservative cover to prevent sunlight from directly reaching the soil (**Figure 12d**).

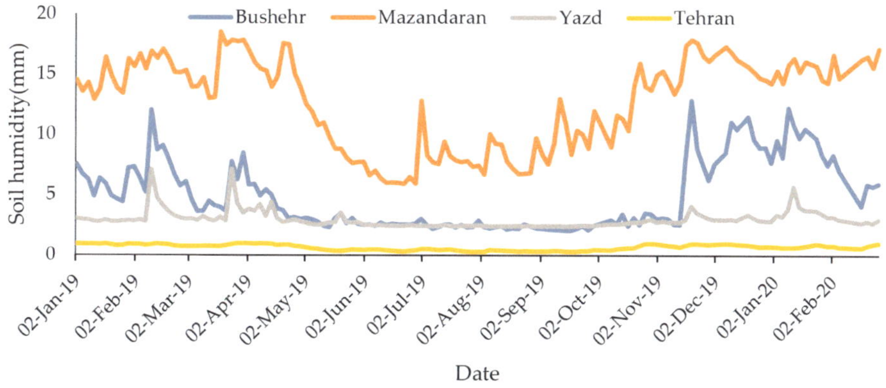

Figure 11.
Soil humidity of different provinces in different seasons.

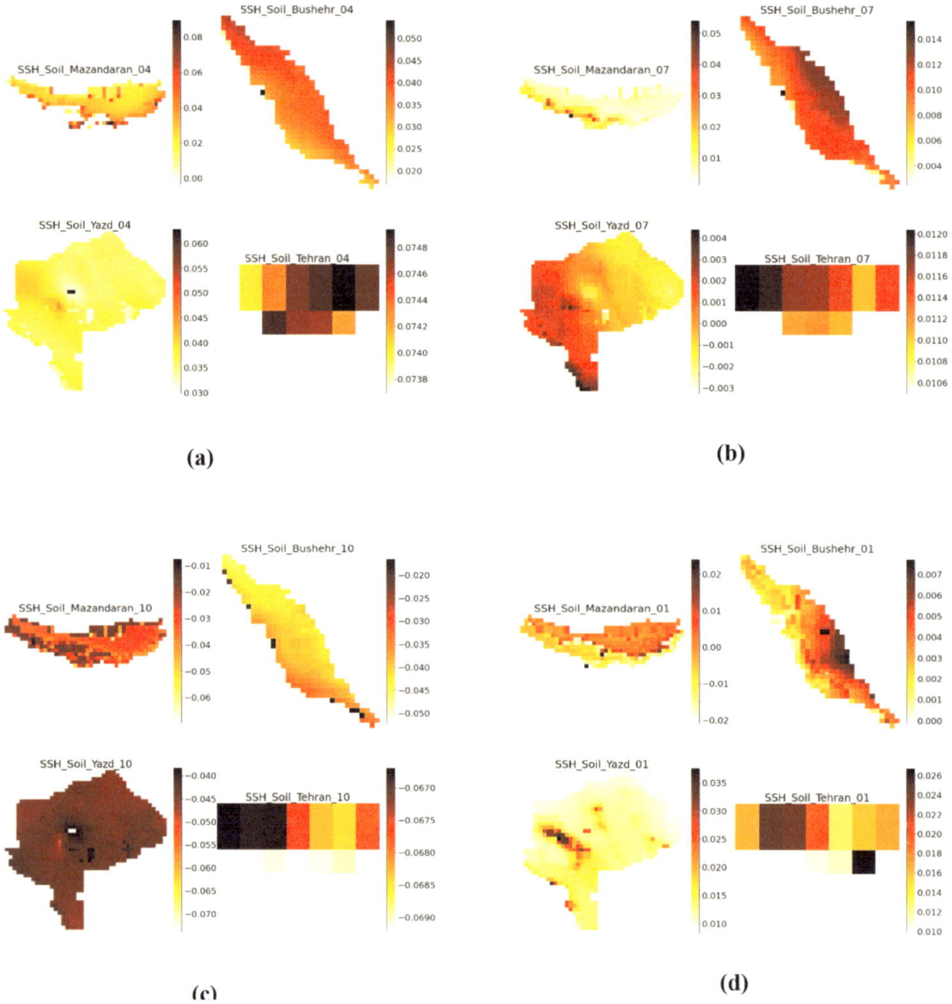

Figure 12.
The entropy of soil heat flux in different climatic zones storage in spring (a), summer (b), autumn (c), and winter (d) (W/m^2).

Sensible heat flux is the rate of energy loss from the soil through convection and diffusion processes as a result of the temperature difference between the surface and the lowest layer of the atmosphere. In the winter season, this difference exists in all regions; therefore, as it is evident from the figures above, the amount of entropy production in the winter season is almost higher in all provinces. on the other hand, the heat flux in the soil is an important part of the energy balance of the crop. Soil acts as a great energy accumulator, which is able to store heat during the day and release it at the night. Something similar happens in annual conditions.

Since the driving force of entropy changes on the soil surface is rainfall, and the surface temperature changes, in areas with dense vegetation cover, the soil is wetter. So, the driving force has the necessary mass transfer to increase the entropy changes caused by heat transfer and phase change. In areas with more rainfall (Mazandaran and Tehran provinces), the soil has produced more soil entropy flux in the spring season, while in the winter and autumn season, Bushehr and Yazd provinces have more entropy changes.

3.5 The entropy of metabolic or biochemical energy storage

Vegetation is an essential component of the surface energy budget, ecosystem performance, and thermodynamic efficiency [25]. As shown in **Figure 13**, the trend of vegetation cover [26, 27] in different provinces has been studied using the MOD13Q1.006 Terra Vegetation Indices 16-Day Global 250 m satellite dataset and the average value of the NDVI index has been calculated for each province for one year.

As shown in **Figure 13**, Mazandaran province mainly has a dense vegetation cover in many seasons, while other provinces have weaker vegetation cover.

The color of objects is obtained from the ratio of absorption to reflection or the passage of radiation waves in the visible range so that if the body absorbs more of the radiation waves, its color will be dark, and in the opposite scenario, its color will appear light. It is the main reason for the entropy generation in autumn.

Plant transpiration in the forest and bare soil land use causes an increase in transpiration heat flux with temperature reduction. So, it plays a critical role in net entropy production on the earth.

In the current study, WAPOR data has been used for satellite imagery data processing. Net primary production (NPP) is studied during a month and the entropy generated in different seasons is determined based on the potential of biomass heat and biochemical energy storage in Gu et al.'s [28] studies. As shown in **Figure 14a–d**, the Bushehr and Mazandaran provinces had a similar procedure in the generation of entropy due to net primary production in winter. In spring and fall, the entropy of metabolic energy storage in Mazandaran is about twice as much as the other provinces except for Yazd province, where it covers bare land use, and season changes do not have any effects on the entropy generated by biochemical energy storage. In mountainous climates, the entropy due to metabolic energy storage has an almost constant trend, which is about half of the wet and cold climate

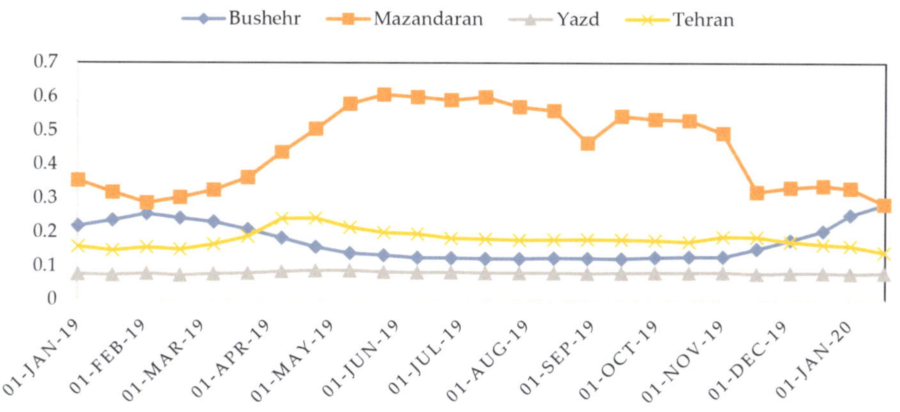

Figure 13.
Vegetation cover changes in case studies (MODIS satellite dataset).

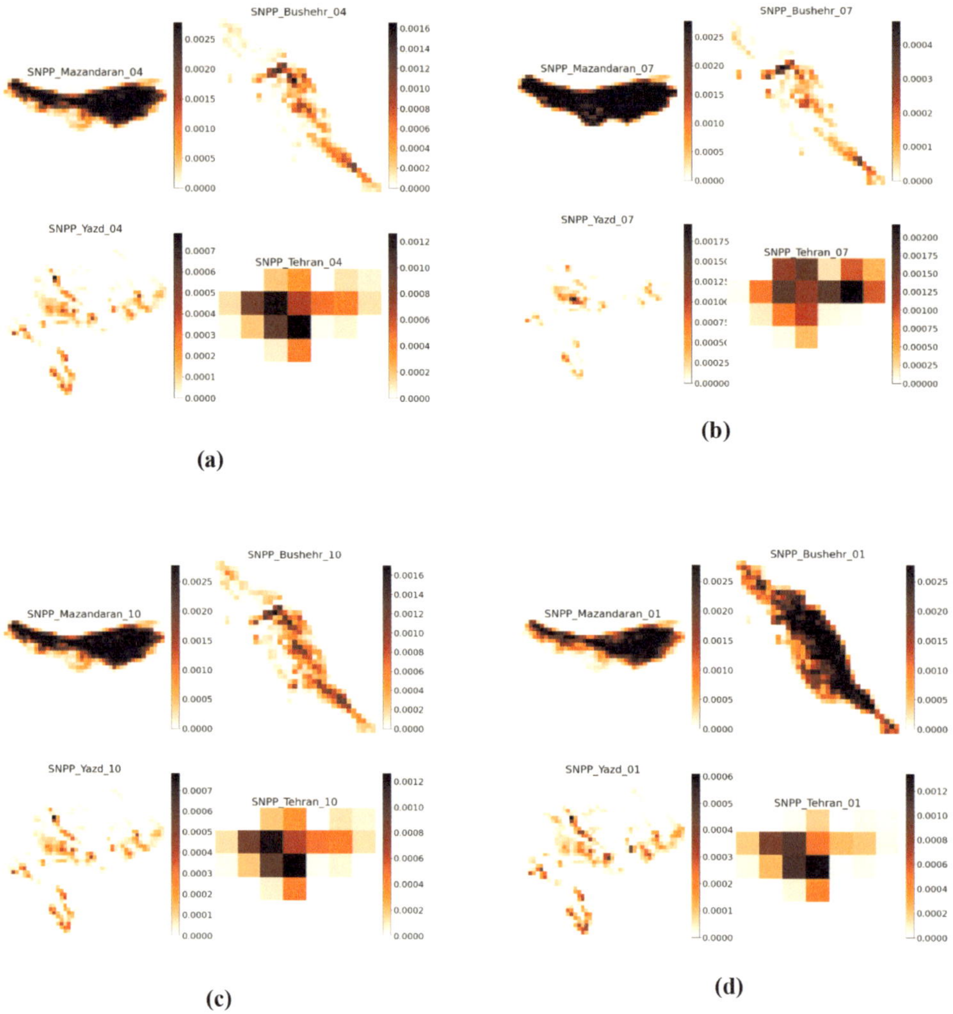

Figure 14.
The entropy of metabolic or biochemical energy storage in spring (a), summer (b), autumn (c), and winter (d) (W/m²).

zone. In summer, it only reaches half of its entropy value in other seasons. This limitation may be created due to water access restrictions, which have reduced agricultural activities.

3.6 Heat storage

The photosynthesis process converts the low-entropy short-wave radiation energy to biochemical energy, which is formed in the vegetation area. Part of this energy is turned into vapor and leaves the plant. This heat energy generates high entropy. It follows a similar trend in biochemical entropy generation (**Figure 15a–d**). It completely depends on the evaporation and transpiration processes of the plant. This is a source of an increase in entropy in the plant. During this process, a phase change

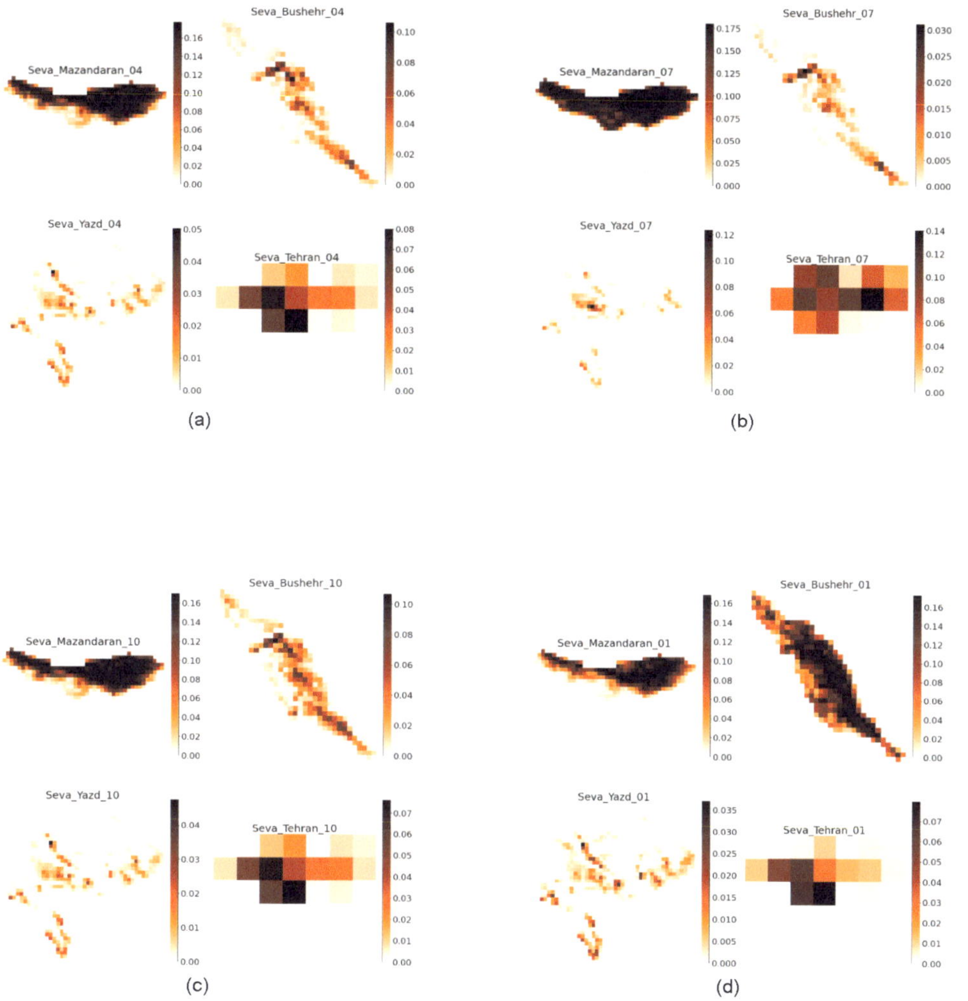

Figure 15.
The entropy of biomass heat storage in spring (21 April–21 May 2019) (a), in summer (23 July–22 August 2019) (b), in autumn (23 October- 21 November 2019) (c), and in winter (21 January-19 February 2019) (d).

occurs, which needs more heat generation. This process leads to an entropy generation level, which is bigger than 10 times of entropy generation in the biochemical energy storage process.

3.7 Entropy generation of different ecosystems in different climatic zones

In spring, we witness a lot of evapotranspiration, latent heat, and soil heat loss in green zones. Mazandaran and Tehran lands are composed of more vegetative surfaces rather than the rest. So, the entropy increment is observable in these two regions (**Figure 16a–d**), which is greater than in other regions. Bushehr has less vegetation land compared to other regions due to the saline quality of its soil. Then, nature will have less reaction to heat generation in this region.

The largest entropy generation occurs in spring. In spring, in cold and wet climates, the temperature is somewhat low due to heavy rainfall, and vegetation growth

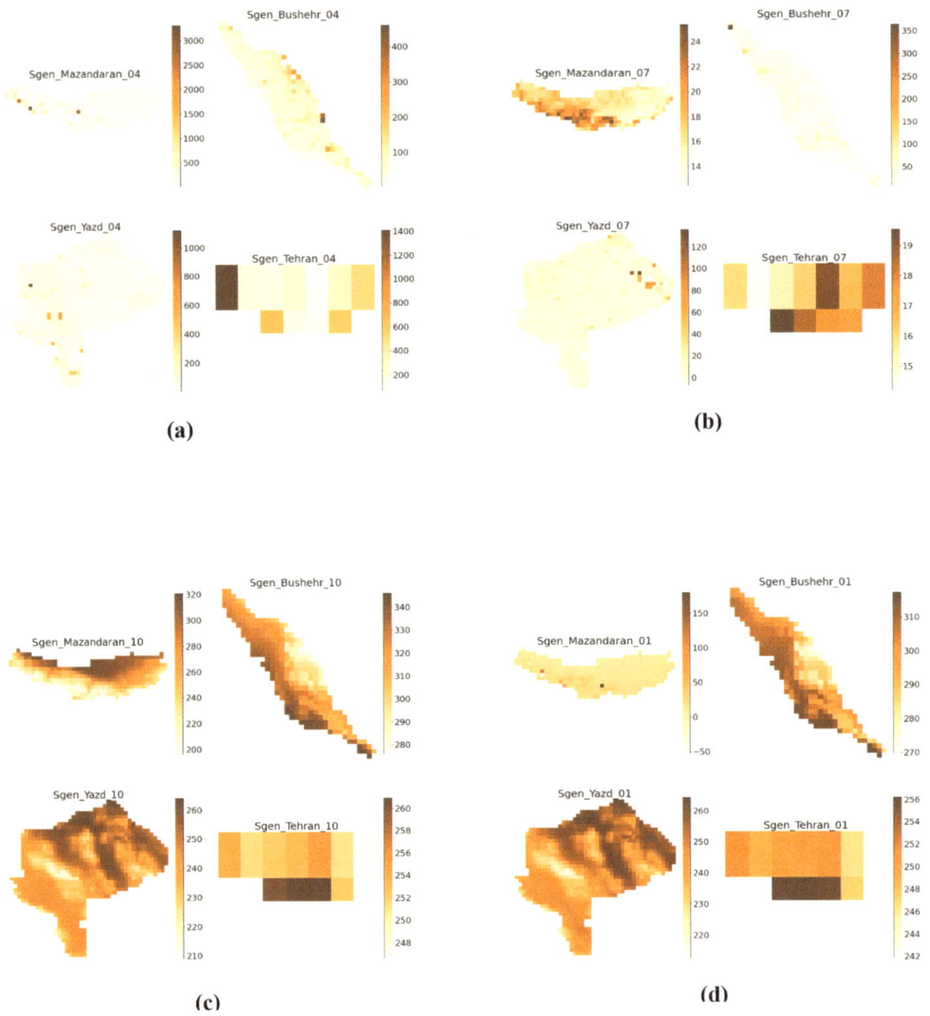

Figure 16.
The entropy generation in spring (21 April–21 May 2019) (a), in summer (23 July-22 August 2019) (b), in autumn (23 October–21 November 2019) (c), and in winter (21 January–19 February 2019) (d).

has a significant effect. Countless evolutions occur in this ecosystem, which causes a significant increase in entropy generation, while in winter, entropy generation of net primary production and evapotranspiration (latent and sensible heat) decreases and so does their influence on the entropy generation. Moreover, in a part of Mazandaran province, negentropy is observed. It shows that nature has made effort to achieve more order in its structure.

Conditions similar to that of Mazandaran province exist around Tehran province. The behavior of nature in this region is considerable due to natural evolution.

Yazd province does not have considerable vegetation cover, and natural evolution is not considered in this region. The main reason for entropy generation in this region is related to the difference between outgoing long-wave entropy flux to space and short-wave absorption. It generates frictional dissipation in the form of entropy produced by adiabatic heating by water phase change (i.e., evaporation and condensation temperature differ), frictional dissipation of falling raindrops, enthalpy transport (horizontal and vertical) by irreversible processes, and energy transport as a result of radiation exchange. The lack of vegetation cover leads to a decrease in the generation of entropy in this region.

In autumn, all regions have the same behavior in entropy generation. In this season, vegetation cover does not change considerably and due to the proximity to the mountain peaks and sea, the temperature decreases, which would become the main cause of entropy generation.

3.8 Carnot efficiency

Carnot efficiency indicates the maximum work of a heat engine. If an ecosystem is considered a heat engine, which has one cold sink (the atmosphere) and one hot source (the sun), it will be able to generate high-quality energy (work) because of the abundant energy of sunlight. This is due to unlimited access to energy (**Figure 17**).

Regarding a different climatic zone, the Carnot efficiency is evaluated in these areas. As can be seen, different ecosystems (different climatic conditions) can produce more than 95% useful work.

The entropy efficiency of the wet and cold climates (**Figure 18a**) is considerably higher than that of the other climates. In this climate, the use of vegetation land encompasses a wider area. It is able to increase the ecosystem's ability to generate work.

Dry and hot climates (**Figure 18c**) include many desert or semi-desert areas, while in mountain climates, rural and urban land use is common. This phenomenon can have a considerable impact on ecosystem performance.

Other climates in some areas have similar behavior to these two climates (**Figure 18b** and **d**).

3.9 The nature efficiency of different climatic zones

In different systems, some losses can be minimized, but as entropy is always increasing they cannot be removed.; that is, the low-entropy heat energy is converted to the high-entropy level because of resource degradation [16]. We are witnessing a

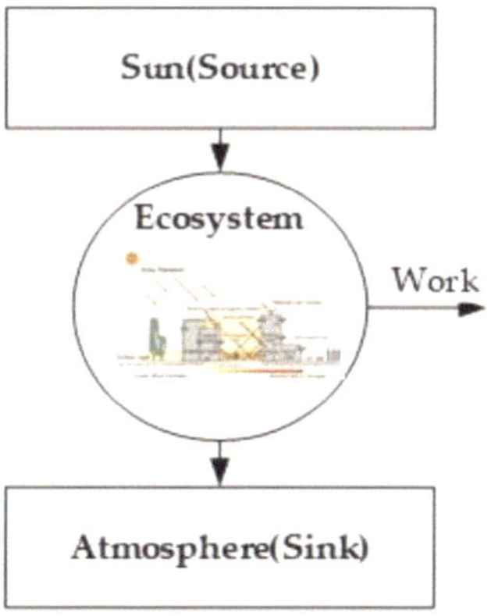

Figure 17.
Carnot efficiency of the ecosystem.

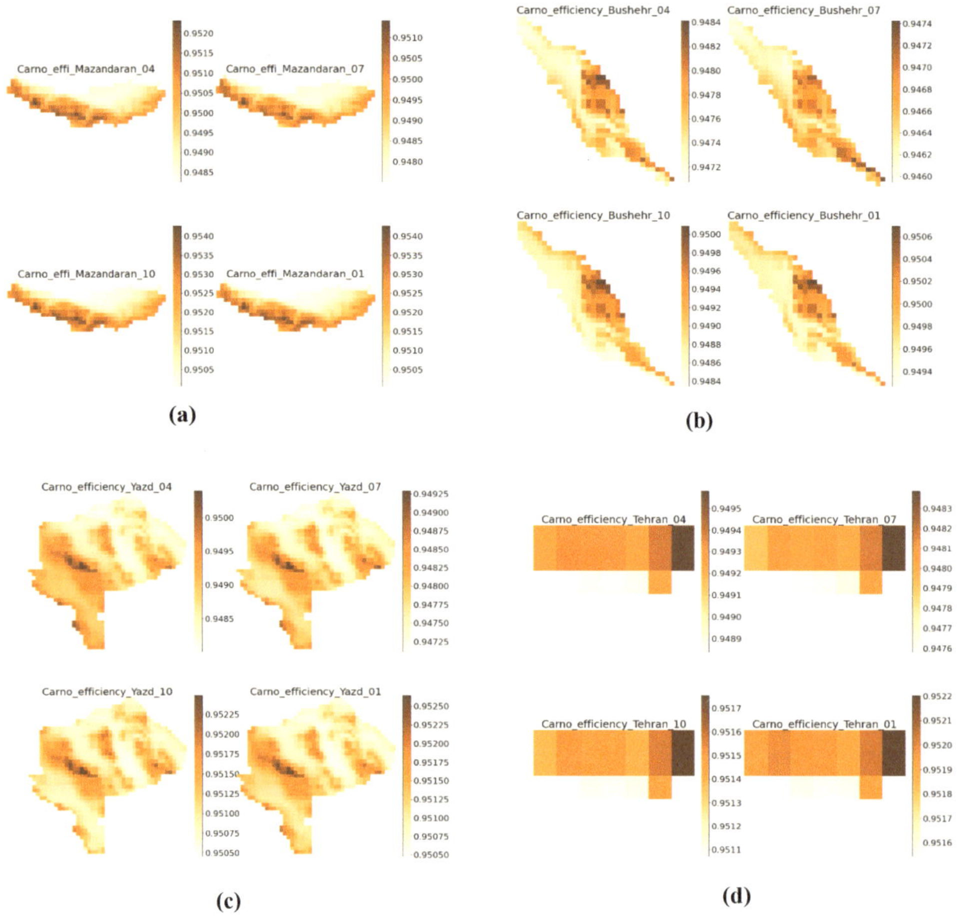

Figure 18.
Carnot efficiency in a wet and cold climate (a), a wet and hot climate (b), a dry and hot climate (c), and a mountain climate (d).

similar system in nature. The Sun provides the possibility of carbon uptake and respiration and maintenance, which can increase ecosystem performance. Low-grade entropy heat energy generates high-entropy heat energy in the ecosystem and it is the origin of countless natural phenomena.

Let us consider a plant and a leaf on it. The leaf absorbs solar energy, and the plant uses heat dissipation to grow more leaves and become a little better at absorbing energy. The energy can be used to dissipate more heat, grow more leaves, and absorb even more energy. It is a positive feedback loop that makes the plant get better and better at dissipating heat. This is a general phenomenon in all matters, living or not. **Figure 19** is a general view of the nature-inspired steam turbine system.

As shown in **Figure 19**, nature is considered a steam generation cycle. A part of solar radiation energy is used for the carbon sequestration process, and the other part as losses is removed from the system. A part of the losses is used in the respiration process (steam turbine) to create growth and maintenance respiration. As a result of this process, a part of the energy is removed from the system as losses. Actually, in

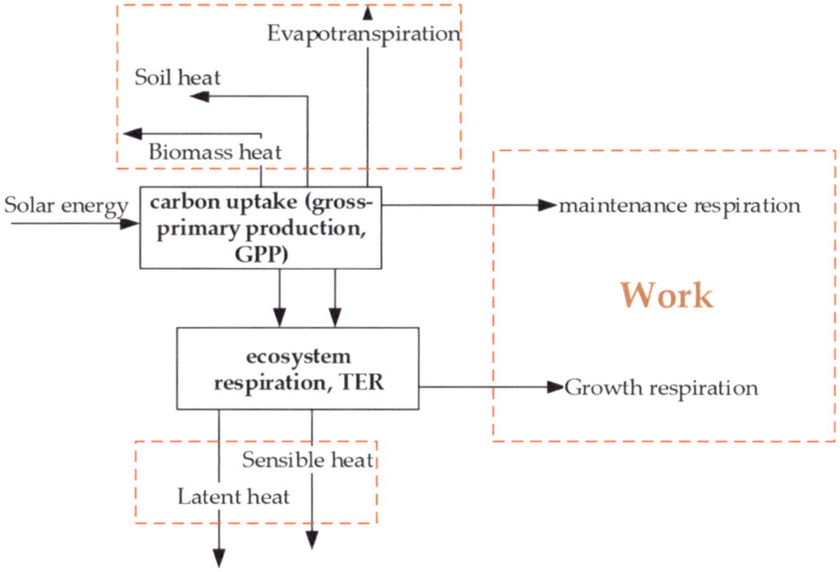

Figure 19.
Ecosystem as a steam power plant.

nature, some of the processes are used to discharge the entropy; that is, depending on the details of the internal processes of the system, the discharge of entropy (condenser in power plant systems) is required, and therefore some of the input energy is removed from the system due to stability requirements.

Living systems are not isolated; that is, they are parts of a larger system. When the entropy generated from energy transformations increases, the energy available for work decreases. In this regard, the purpose of this part is to investigate the effect of the temperature changes arising from season variations on entropy generation in different climatic zones. It is clear that human activities should be considered in this study. Here, different flows in nature are considered.

Mazandaran province is one of the greenest areas of Iran where intensive agricultural activities have always been done there. As can be seen in **Figure 20a**, in areas close to the sea, located in the north of Mazandaran province, due to tourism development, Urban land use has been predominant in this area, while in the southern areas of Mazandaran province, there have been many forests, and large amounts of crops have been cultivated in these areas. So, the cultivation in these areas is somewhat pristine, and as a result, the earth can breathe easily there.

Due to the economic activities in the south of Bushehr, the entropy efficiency has had the lowest value in that area (south of Bushehr). On the contrary, in the northern areas of Bushehr, the land has great potential for the cultivation of various crops. Moreover, there are many forests in these areas, which are the main sources of organic carbon generation in the soil. Therefore, nature is more inclined to generate power in these areas (**Figure 20b**).

The quality in Yazd has a relatively uniform distribution. Many crops are harvested in spring and fall in Yazd. Therefore, nature is more able to use the input energy of the sunlight during these seasons (**Figure 20c**).

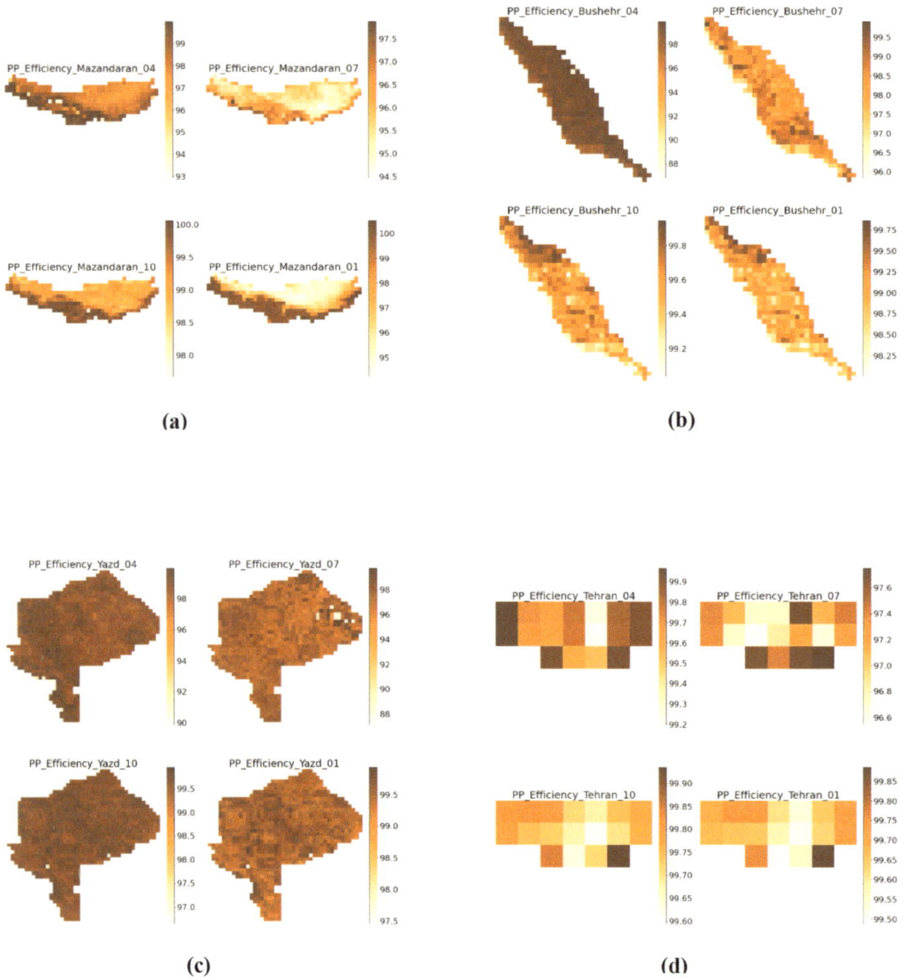

Figure 20.
Entropy efficiency of nature in a wet and cold climate (a), a wet and hot climate (b), a dry and hot climate (c), and a mountain climate (d).

Figure 20d illustrates this in different seasons. In the center of Tehran, we are witnessing a lot of anthropological activities; while in the countryside of Tehran particularly near the south and west sides of Tehran, nature has the potential needed for agricultural activities.

4. Conclusions

In spring, due to the growth of plants, chemical reactions lead to entropy generation. This behavior is more in Mazandaran province due to the high level of vegetation area regarding its climate. In the winter season, this province has the minimum growth rate and therefore has less entropy production. Other provinces have grown mildly due to the type of climate. So, they have a greater increase in the entropy generation than Mazandaran province. In the autumn season, due to the change in receiving light wavelengths, chlorophyll reactions are not performed, therefore, the amount of entropy production in this area is also lower than in the spring season. In the summer season, due to the high sunlight absorption, plant growth decreases in many northern regions such as Mazandaran and Tehran proviences, so entropy production in this time zone has reduced. This is even though in the winter season, as the location of Yazd and Bushehr in the southern region, the absorption of sunlight is suitable for plant growth in these areas, and therefore, entropy generation increases in these areas in the winter season.

Moreover, in the spring season, evapotranspiration occurs in areas with more vegetation cover. So, the amount of water generation will be higher in these areas. The latent heat content of water vapor in terms of entropy generation is greater in these areas. In other words, the behavior of biological growth is effective on the rate of evapotranspiration, which leads to an increase in the entropy generated due to the latent heat process.

Considering that all the processes in nature are very slow, the variation of entropy generation is not very noticeable. So, the efficiency of nature, and the natural Carnot efficiency are very close. This means that despite the changes in the generated entropy caused by the processes, the self-organization of nature has occurred in terms of negentropy, which leads to increases the nature's efficiency. Since human activities are not included in this study, it is assumed that only self-organization has happened in different areas. Otherwise, it will create a problem and reduce the yield of nature and the yield of nature's life.

Nature is always evolving. These changes seem to be systemized via interaction with different biotic and abiotic activities. The only important principle is to reach sustainability, how we shall find the natural approach in the energy systems. Nature reacts to various side effects of anthropological activities. This process is everlasting. The more the knowledge about nature, the more the understanding about the end of life on the earth.

Data availability statement

All of the chapter's data has been provided by the Satelite dataset.

Author details

Masoumeh Bararzadeh Ledari[1]* and Reza Bararzadeh Ledari[2]

1 Department of Energy Engineering, Sharif University of Technology, Tehran, Iran

2 Farab Company, Tehran, Iran

*Address all correspondence to: bararzadeh@energy.sharif.ir

References

[1] Pettorelli N. Climate change as a main driver of ecological research. Journal of Applied Ecology. 2012;**49**(3): 542-545. DOI: 10.1111/j.1365-2664. 2012.02146.x

[2] Buchanan M. The thermodynamics of Earth. Nature Physics [Internet]. 1 Feb 2017;**13**(2):106. Available from: http:// www.nature.com/articles/nphys4031

[3] Hermann WA. Quantifying global exergy resources. Energy. 2006;**31**(12): 1349-1366. DOI: 10.1016/j.energy.2005. 09.006

[4] Lovelock J. The Ages of Gaia: A Biography of Our Living Earth. illustrate. Oxford University Press; 1995

[5] Fraser R, Luvall JC, Ulanowicz RE. Can we Use Energy Based Indicators to Characterize and Measure the Status of Can we Use Energy Based Indicators to Characterize and Measure the Status of Ecosystems, Human, Disturbed and Natural? In: Advances in Energy Studies: Exploring Supplies, Constraints and Strate ies. Service Grafici Editoriali; 2001

[6] Stoy PC. Thermodynamic approaches to ecosystem behaviour: fundamental principles with case studies from forest succession and management. In: Raffaelli DG, Frid CLJE, editors. Ecosystem Ecology [Internet]. Cambridge University Press; 2010. pp. 40–64. (Ecological Reviews). Available from: https://www.cambridge.org/core/ product/identifier/CBO978051175045 8A010/type/book_part

[7] Lin H, Cao M, Stoy PC, Zhang Y. Assessing self-organization of plant communities—A thermodynamic approach. Ecological Modelling. 2009; **220**(6):784-790. DOI: 10.1016/j.ecol model.2009.01.003

[8] Lin H, Cao M, Zhang Y. Self-organization of tropical seasonal rain forest in Southwest China. Ecological Modelling. 2011;**222**(15):2812-2816. DOI: 10.1016/j. ecolmodel.2010.07.006

[9] Schneider ED, Kay JJ. Life as a manifestation of the 2nd law of thermodynamics. Mathematical and Computer Modelling. 1994;**19**(6): 25-48. DOI: 10.1016/0895-7177(94) 90188-0

[10] Steinborn W, Svirezhev Y. Entropy as an indicator of sustainability in agro-ecosystems: North Germany case study. Ecol Modell. 2000;**133**:247-257. DOI: 10.1016/S0304-3800(00)00323-9

[11] Lovelock JE. Gaia: A New Look at Life on Earth. illustrate. Oxford University Press; 1979. p. 864

[12] Ledari MB, Saboohi Y, Valero A, Azamian S. Exergy cost analysis of soil-plant system. International Journal of Exergy. 2022;**38**(3):293. DOI: 10.1504/ IJEX.2022.124174

[13] Gan Z, Yan Y, Qi Y. Entropy budget of the earth, atmosphere and ocean system. Progress in Natural Science. 2004;**14**:1088-1094. DOI: 10.1080/ 10020070412331344851

[14] Agalhães G. Some Reflections on Life and Physics: Negentropy and Eurhythmy. Quantum Matter [Internet]. 1 Jun 2015;**4**(3):258–266. Available from: http://openurl.ingenta.com/content/ xref?genre=article&issn=2164-7615& volume=4&issue=3&spage=258

[15] Norris C, Hobson P, Ibisch PL. Microclimate and vegetation function as indicators of forest thermodynamic efficiency. Journal Application of

Ecology. 2011. DOI: 10.1111/j.1365-2664.2011.02084.x

[16] Ibrahim Dincer, and Y. A. C. Energy, entropy and exergy concepts and their roles in thermal engineering. Entropy. 2001;**3**:116-149

[17] Peixoto JP, Oort AH, De Almeida M, Tome A. Entropy budget of the atmosphere. Journal of Geophysical Research. 1991;**96**(D6):981-988. DOI: 10.1029/91jd00721

[18] Wu XF, Chen GQ, Wu XD, Yang Q, Alsaedi A, Hayat T, et al. Renewability and sustainability of biogas system: Cosmic exergy based assessment for a case in China. Renewable and Sustainable Energy Reviews. 2015;**51**:1509-1524. DOI: 10.1016/j.rser.2015.07.051

[19] Bararzadeh Ledari M, Saboohi Y, Valero A, Azamian S. Exergy analysis of a bio-system: Soil–plant interaction. Entropy. 2021;**23**(1). DOI: 10.3390/e23010003

[20] Majozi NP, Mannaerts CM, Ramoelo A, Mathieu R, Nickless A, W. V. Analysing surface energy balance closure and partitioning over a semi-arid savanna FLUXNET site in Skukuza, Kruger National Park, South Africa. Hydrological Earth System Science. 2017;**21**:3401-3415. DOI: 10.5194/hess-21-3401-2017

[21] Holdaway RJ, Sparrow AD, Coomes DA. Trends in entropy production during ecosystem development in the Amazon Basin. Philosophical Transactions of the Royal Society B, Biological Sciences [Internet]. 12 May 2010;**365**(1545):1437–1447. Available from: https://royalsocietypublishing.org/doi/10.1098/rstb.2009.0298

[22] Brunsell NA, Schymanski SJ, Kleidon A. Quantifying the thermodynamic entropy budget of the land surface: Is this useful? Earth System Dynamics. 2011;**2**:71-103. DOI: 10.5194/esdd-2-71-2011

[23] Hejazizadeh Z, Bazmi N, Alireza Rahimi MTN. Spatial-temporal modeling of albedo in Iran. Journal of Applied Research in Geographical Sciences. 2018; **47**(17):1-17

[24] Han C, Ma Y, Chen X, Su Z. Estimates of land surface heat fluxes of the Mt. Everest region over the Tibetan Plateau utilizing ASTER data. Atmospheric Research. 2016;**168**:180-190. DOI: 10.1016/j.atmosres.2015.09.012

[25] Zhang H, Wu J. A statistical thermodynamic model of the organizational order of vegetation. Ecological Modelling. 2002;**153**(1): 69-80. DOI: 10.1016/S0304-3800(01)00502-6

[26] Matsushita B, Yang W, Chen J, Onda Y, Qiu G. Sensitivity of the Enhanced Vegetation Index (EVI) and Normalized Difference Vegetation Index (NDVI) to topographic effects: A case study in high-density cypress forest. Sensors. 2007;**7**(11):2636-2651. DOI: 10.3390/s7112636

[27] Pettorelli N, Vik JO, Mysterud A, Gaillard JM, Tucker CJ, Stenseth NC. Using the satellite-derived NDVI to assess ecological responses to environmental change. Trends in Ecology and Evolution. 2005;**20**(9): 503-510. DOI: 10.1016/j.tree.2005.05.011

[28] Gu L, Meyers T, Pallardy SG, Hanson PJ, Yang B, Heuer M, et al. Influence of biomass heat and biochemical energy storages on the land surface fluxes and radiative temperature. Journal of Geophysical Research Atmospheres. 2007;**112**(2):1-11. DOI: 10.1029/2006JD007425

Chapter 7

State-of-the-Art on the Marine Current Turbine System Faults

Sana Toumi, Mohamed Benbouzid and Mohamed Faouzi Mimouni

Abstract

This chapter deals with the state of the art on the marine current turbine (MCT) system faults. Indeed, the MCT structure consists of a marine turbine, a generator (permanent magnet synchronous generator (PMSG) or doubly fed induction generator (DFIG)), and a PWM power converter. Nevertheless, these systems are exposed to functional and environmental severe conditions. Firstly, the power increase leads to a higher current and/or voltage. Second, the installation of the MCT system under the sea and the existence of the swell and wave imply harmonic current speeds. In fact, several faults (related to the turbine, the generator, the blades, and the converters) can occur in the MCT system. Most of these faults generate the speed and the torque oscillations, which can lead to mechanical vibrations and the rapid destruction of the insulating material generator. Consequently, MCT system performances can be degraded.

Keywords: marine current turbine, permanent magnet synchronous generator, doubly fed induction generator, converter, faults

1. Introduction

Today, the use of the marine current turbine, shown in **Figure 1**, has a great importance in the electrical energy production. Nevertheless, these systems are exposed to functional and environmental severe conditions. Indeed, different faults can exist in these systems, whether on the turbine marine, on the generator, or on the converters [1]. Therefore, the detection of the MCT faults becomes essential in order to minimize the maintenance cost and ensure the continuity of the electricity production.

Moreover, permanent magnet synchronous machines or doubly fed induction generators can be used for MCT systems [2, 3]. Thus, the existence of any fault can lead to an abnormal behavior of the machine and its accelerated aging process [4]. The main faults of permanent magnet synchronous machines can be summarized in two main categories; stator faults (short-circuit between turns, short circuit between phase and neutral, and short-circuit between phases) and rotor faults (magnets (for the PMSG), eccentricity, and rotor windings faults) [5].

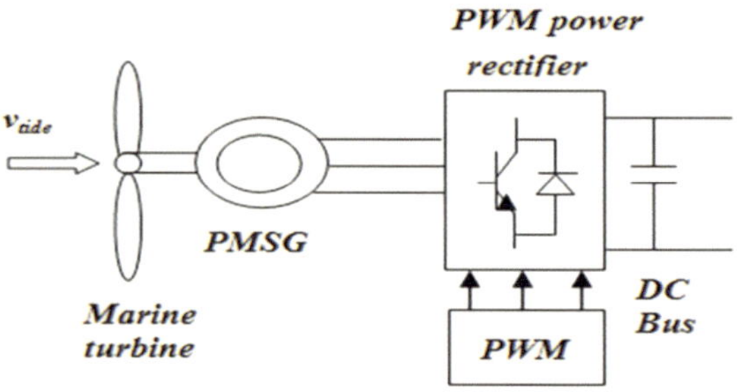

Figure 1.
An off-grid marine current turbine structure.

Furthermore, various studies have presented that 70% of converter faults are related to power switches [6, 7]. Indeed, insulated gate bipolar transistors (IGBTs) are rugged, but, because of electrical stress and excess thermal, they suffer from failure. Converters faults can be, generally, divided into intermittent gate misfiring faults, open-circuit faults, and short-circuit faults. Thus, these faults must be taken into account; otherwise, they can lead to the whole system's performance degradation [8, 9].

This chapter describes the different faults that can be occurring in the marine current turbine system. It is composed as follows; in Section 2, marine turbine faults are described. In Section 3, generator faults and their factors are represented. In Section 4, different types of converter faults are described. The conclusion is given in Section 5.

2. Marine turbine faults

Marine turbine faults are mainly located at the blades. Indeed, seawater salinity causes the corrosion and the rusting of the blades (**Figure 2**). This requires regular maintenance and the use of a high-performance anti-rust coating. Also, the presence of algae around the blades can also damage the marine turbine.

Figure 2.
Marine turbine fault.

3. Generator faults

For MCT configuration, both the permanent magnet synchronous generator and the doubly-fed induction generator have been used. **Table 1** gives the advantages and disadvantages of every generator.

3.1 Stator faults

Stator faults are due to different factors.

- Thermal factors

For a nominal temperature, the insulating material that covers conductors has a well-defined lifespan. Because of a voltage variation or a repetition of starts in a very short time, the temperature can increase and exceed the nominal operating temperature, which leads to the insulating material life reduction [10].

- Electrical factors

The dielectric properties of the insulating material can be contaminated by foreign things (fats, dust, ...), which can cause a small current discharge that leads to a short circuit between conductors and magnetic carcasses.

- Mechanical origins

Repetitive starts of the machine cause the temperature rise, which leads to the insulation dilation. This could generate breaks in the insulation that implies a short circuit fault [10].

- Environmental origin

In general, humidity and the presence of chemicals in ambient air can degrade the insulating material quality and affect its lifespan [11].

The main faults of the stator are given by (**Figure 3**); short-circuit between turns (a), short circuit between phase and neutral (b), and short-circuit between phases (c) [12]. These faults generate a disturbance in the spatial distribution of the rotating

Type	Advantages	Disadvantages
PMSG	• Full speed range • Possibility to avoid gearbox (direct drive) • Complete control of reactive and active power	• Full-scale power converter • Low-speed generator (big and heavy) • Permanent magnets needed
DFIG	• Limited speed range (±30% around synchronous speed) • Low-cost small capacity PWM inverter • Complete control of reactive and active power	• Need slip rings • Need for gear

Table 1.
Generator topologies comparison.

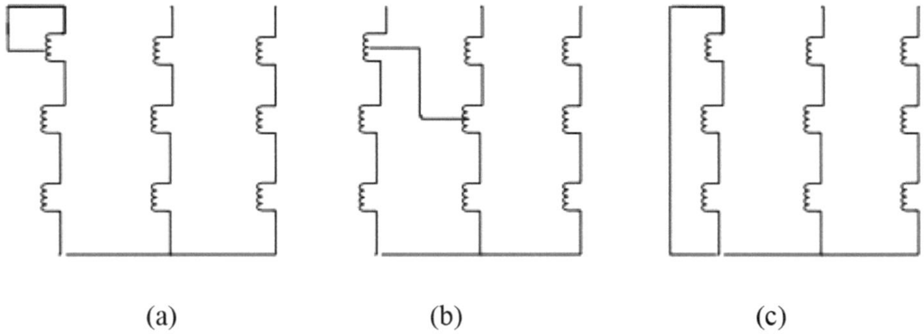

Figure 3.
Different short-circuit faults in the stator.

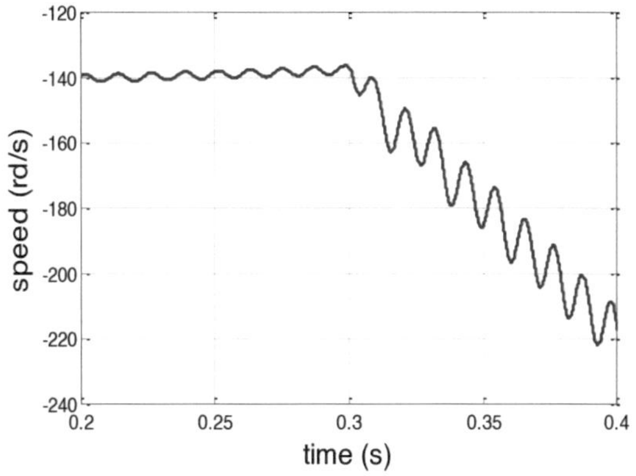

Figure 4.
Speed curve before and after fault.

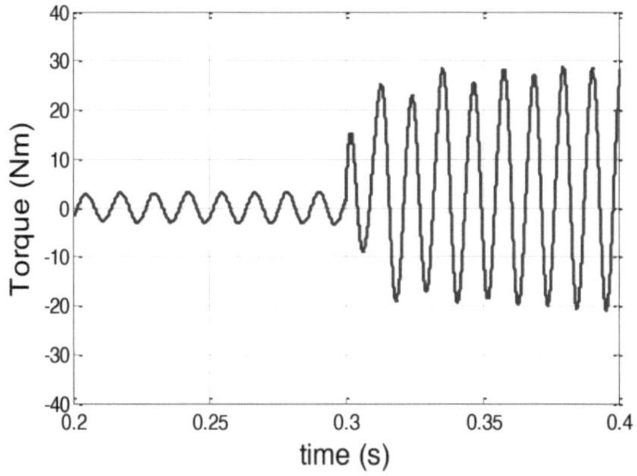

Figure 5.
Torque curve before and after fault.

field. This causes, first of all, the electromagnetic torque and the speed oscillations (**Figures 4** and **5**, respectively), which lead to mechanical vibrations. Moreover, the short-circuit current with important values can lead to the rapid destruction of the insulating material (**Figure 6**) [13, 14].

Figure 6.
Insulating material destruction.

3.2 Rotor faults

- Rotor winding faults

These faults present between 42 and 50% of all rotor failures in electrical machines. It can generate noise, mechanical vibrations, and the bearing wear rise. Moreover, winding faults are 12 times more frequent in inverter-powered machines than in those powered directly by the grid [14].

- Eccentricity

The stator and the rotor, for an electrical machine in healthy conditions, have the same rotation axis (same center) (**Figure 7**). Eccentricity is defined as an unsymmetrical air gap between the rotor and the stator. It can be static or dynamic.

For the static eccentricity, the stator and the rotor centers are different (**Figure 8**), however, the rotor always turns around its axis. The main cause of static eccentricity is the incorrect position of the stator and the rotor [15].

In the case of dynamic eccentricity, the rotor does not turn around its axis of rotation where there is an unbalance as shown in **Figure 9**. The main cause of the dynamic eccentricity is the mechanical resonances at critical speed [16] that leads to the electromagnetic forces increase.

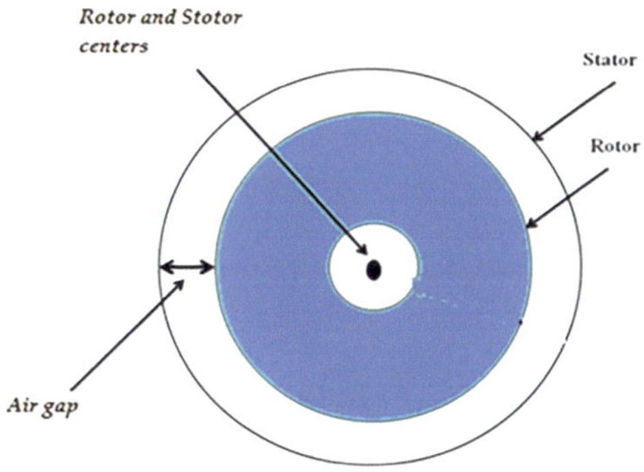

Figure 7.
Stator and rotor in healthy conditions.

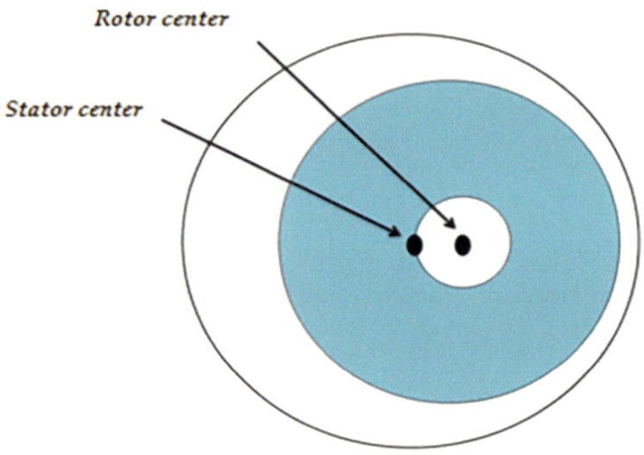

Figure 8.
Static eccentricity.

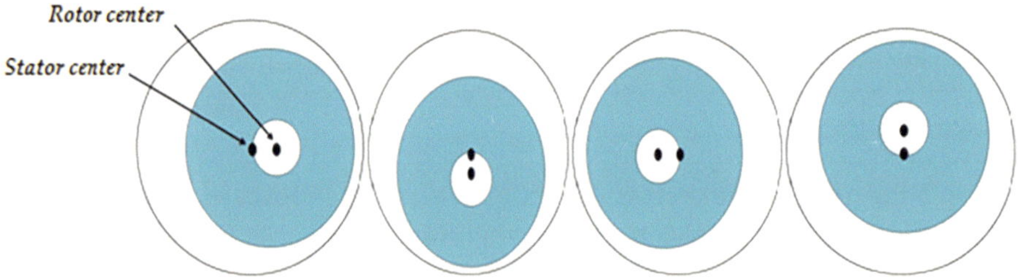

Figure 9.
Dynamic eccentricity.

4. Converter faults

As shown in **Figure 10**, the MCT converter consists of three legs. Each leg is composed of two semiconductor switches (T_k, Tk_{+3} k = 1, 2, 3) and two freewheeling diodes (D_k, Dk_{+3}). For the switches control, a PWM bloc based on logic control signals S_k (k = 1, 2, 3) is used, and it is expressed by:

$$S_k = \begin{cases} 1\,if\ T_k on\ and\ T_{k+3}\ off \\ 0\,if\ T_{k+3} on\ and\ T_k\ off \end{cases} \tag{1}$$

Most converter faults are short-circuit and open-circuit faults.

4.1 Open-circuit fault

IGBT's open-circuit faults are, generally, caused by the loss of bonding wires of the control signal or the transistor rupture by a short-circuit [17]. Furthermore, this fault can occur when the switches are destructed by an accidental overcurrent or a fuse connected with series for short protection is blown out. It can arise on the upper switch (**Figure 11a**), the lower switch (**Figure 11b**), or even the two switches at the same time (**Figure 11c**).

Following this fault, the converter cannot synthesize balanced output voltages, thus providing large torque ripple (**Figure 12**) and increasing speed harmonics distortion (**Figure 13**) [18].

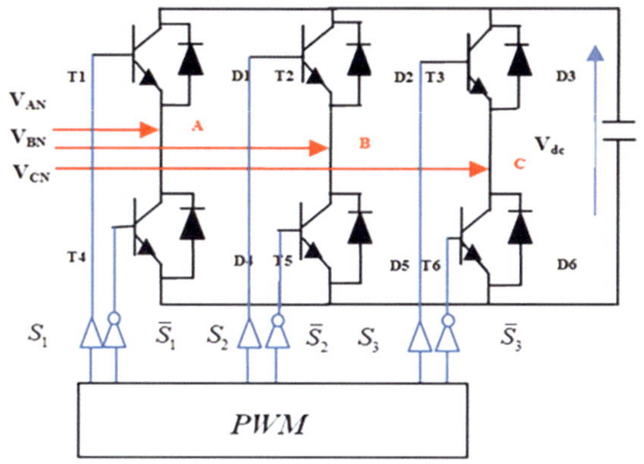

Figure 10.
Converter topology.

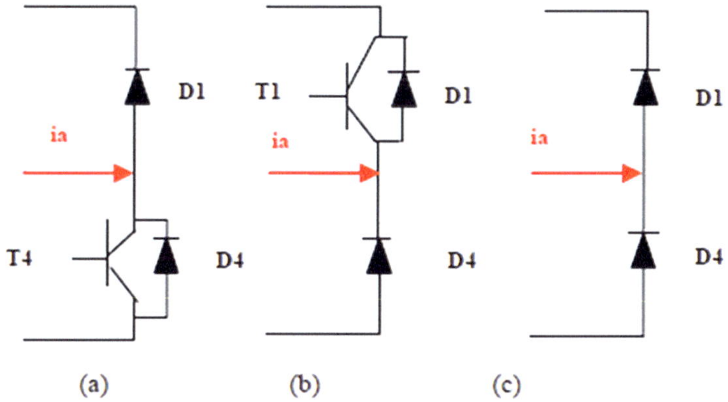

Figure 11.
Open-circuit fault.

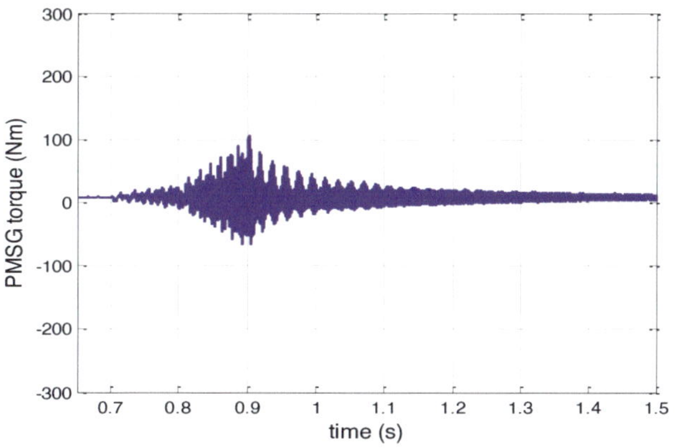

Figure 12.
Torque curve.

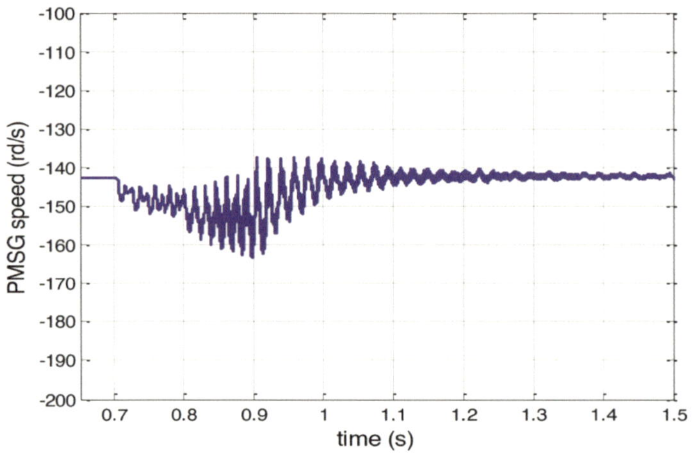

Figure 13.
Speed curve.

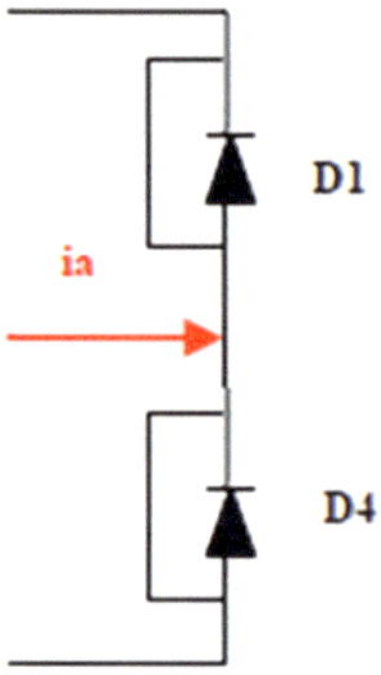

Figure 14.
Short-circuit fault.

4.2 Short-circuit fault

In this case, the fault transistor current increases. When this fault is applied to the whole leg, this latter becomes definitively short-circuited (**Figure 14**). The phase currents become strongly unbalanced and their amplitudes can reach several times that of the currents in normal operation. This not only causes very high torque ripples but can also damage other converter components. In addition, the short-circuit current can result in significant amplitudes.

5. Conclusion

This chapter presents the different faults that can occur in the marine current turbine system, either on the turbine, the machine, or the converter. In this contest, these faults have been analyzed and described by giving each one its impact on the MCT behavior. In fact, Currents, voltages, torque, and speed have been attacked by strong ripples, which lead to the degradation of the MCT performances. Thus, the above-presented study should be useful for the design of advanced robust control techniques in order to correct the effects of the faults and improve the MCT performances.

Author details

Sana Toumi[1,2]*, Mohamed Benbouzid[2,3] and Mohamed Faouzi Mimouni[1]

1 Laboratory of Automation, Electrical Systems and Environment (LASEE), University of Monastir, Monastir, Tunisia

2 University of Brest, Brest, France

3 Shanghai Maritime University, Shanghai, China

*Address all correspondence to: sanaatoumii@gmail.com

References

[1] Toumi S, Elbouchikhi E, Amirat Y, Benbouzid MEH, Feld G. Magnet failure-resilient control of a direct-drive tidal turbine. Ocean Engineering. 2019;**187**:106207

[2] Zhou Z, Scuiller F, Charpentier JF, Benbouzid MEH, Tang T. Grid-connected marine current generation system power smoothing control using supercapacitors. In: Proceedings of the IEEE Annual Conference on IEEE Industrial Electronics Society (IECON'38); 25-28 October 2012. Montreal, Canada: IEEE; 2012. pp. 1-6

[3] Bachir S, Tnani S, Trigeassou JC, Champenois G. Diagnosis by parameter estimation of stator and rotor faults occurring in induction machines. IEEE Transactions on Industrial Electronics. 2006;**53**:963-973

[4] Awadallah MA, Morcos MM, Gopalakrishnan S, Nehl TW. Detection of stator short circuits in VSI-Fed Brushless DC motors using wavelet transform. IEEE Transactions on Energy Conversion. 2006;**21**:1-8

[5] Toumi S, Amirat Y, Elbouchikhi E, Trabelsi M, Benbouzid MEH, Mimouni MF. A simplified mathematical approach for magnet defects modeling in PMSG-based marine current turbine. In: Proceedings of the IEEE International Conference on Sciences and Techniques of Automatic Control and and Computer Engineering (STA'17); 19-21 December 2016. Sousse, Tunisia: IEEE; 2016. pp. 552-557

[6] Toumi S, Amirat Y, Elbouchikhi E, Trabelsi M, Benbouzid MEH, Mimouni MF. A comparison of fault-tolerant control strategies for a PMSG-based marine current turbine system under converter faulty conditions. Journal of Electrical Systems. 2017;**13**:472-488

[7] Toumi S, Amirat Y, Trabelsi M, Elbouchikhi E, Mimouni MF, Benbouzid MEH. Backstepping control of a PMSG-based marine current turbine system under faulty conditions. In: Proceedings of the IEEE International Renewable Energy Congres (IREC'09); 20-22 March 2018. Hammamet, Tunisia: IEEE; 2018. pp. 1-6

[8] Amirat Y, Benbouzid MEH, AlAhmer E, Bensaker B, Turri S. A brief status on condition monitoring and fault diagnosis in wind energy conversion systems. Renewable and Sustainable Energy Reviews. 2009;**3**:2629-2636

[9] Parker MA, Ran L, Finney SJ. Distributed control of a fault-tolerant modular multilevel inverter for direct-drive wind turbine grid interfacing. IEEE Transactions on Industrial Electronics. 2013;**2013**:60

[10] Zarko D, Ban D, Lipo TA. Analytical solution for cogging torque in surface permanent-magnet motors using conformal mapping. IEEE Transactions on Magnetics. 2008;**44**:52-65

[11] Thomson JS, Kallesoe CS. Stator fault modelling of induction motors. In: Proceedings of the IEEE International Symposium on Power Electronics, Electrical Drives, Automation and Motion (SPEEDAM); 23-26 2006. Taormina, Italy: IEEE; 2006

[12] Toumi S, Benelghali S, Trabelsi M, Elbouchikhi E, Amirat Y, Benbouzid MEH, et al. Modeling and simulation of a PMSG-based marine current turbine system under faulty rectifier conditions. Electric Power Components & Systems. 2017;**45**:715-725

[13] Tallam RM, Lee SB, Stone G, Kliman GB, Yoo J, Habetler TG, et al.

A survey of methods for detection of stator-related faults in induction machines. IEEE Transactions on Indudtry Applications. 2007;**43**:920-933

[14] Kliman GB, Lee SB, Shah MR, Lusted RM, Nair NK. New method for synchronous generator core quality evaluation. IEEE Transactions on Energy Conversion. 2004;**19**:576-582

[15] Devanneaux V, Kabbaj H, Dagues B, Faucher J. An accurate model of squirrel cage induction machines under static, dynamic or mixed eccentricity. In: Proceedings of the IEEE International Symposium on Diagnostics for Electrical Machine, Power Electronics and Drives (SDEMPED); 01-03 September 2001. Grado, Italy: IEEE; 2001. pp. 121-126

[16] Andriamalala RN, Razik H, Baghli L, Sargos FM. Eccentricity fault diagnosis of a dual-stator winding induction machine drive considering the slotting effects. IEEE Transactions on Industrial Electronics. 2008;**55**:4238-4251

[17] Lu B, Sharma SK. A literature review of IGBT fault diagnostic and protection methods for power inverters. IEEE Transactions on Industry Applications. 2009;**45**:1770-1777

[18] Toumi S, Benelghali S, Trabelsi M, Elbouchikhi E, Benbouzid MEH, Mimouni MF. Robustness analysis and evaluation of a PMSG-based marine current turbine system under faulty conditions. In: Proceedings of the IEEE International Conference on Sciences and Techniques of Automatic Control and Computer Engineering (STA'15); 21-23 December 2014. Hammamet, Tunisia: IEEE; 2014. pp. 631-636

Chapter 8

Using Exergy-Based Metrics in Assessing Sustainability of Fossil-Fueled Thermal Energy Systems

Ismaila Badmus

Abstract

This chapter examines the importance of exergy-based parameters like exergy efficiency, environmental compatibility, sustainability index, depletion number, and improvement potential of hydrocarbon fuel utilization. The main import of system exergy efficiency is relatively well-known. A hydrocarbon fuel environmental compatibility (ζ) evaluates the fuel exergy performance when its combustion gases emission abatement exergy is factored in. A fuel with low emission abatement exergy has a high environmental compatibility and, thus, high sustainability. Another metric is the depletion number, D_p. This measures the rate of fuel exergy destruction with respect to the fuel input exergy. Since fuel exergy flow is directly related to its material flow, its exergy destruction is similarly directly related to its material depletion. Hence, fuel utilization sustainability necessitates a low D_p. D_p indicates the fraction of input energy resources degraded through entropy creation, turning them into thermodynamic states of no useful energy values. The sustainability index is the reciprocal of D_p. The Improvement Potential (IP) is, mathematically, the product of the square of D_p and the fuel input exergy. When IP is high, it means the exergy losses are too high and there is a big room for exergy efficiency improvement.

Keywords: sustainability index, exergy, metrics, fossil fuel, environmental compatibility, thermal energy, depletion number, improvement potential

1. Introduction

Every activity in the cosmos, anywhere, anytime, involves energy utilzsation through its transformation. This transformation process is guided and dictated by relevant natural laws. The most important and relevant natural laws in energy transformation are the first and second laws of thermodynamics. While the first law stipulates energy quantity conservation, the second law draws attention to the fact that energy types are in different quality grades and that, aside from photosynthesis, energy transformation processes lead to energy quality degradation. Exergy is a concept that facilitates the ranking of energy types based on their quality grades, as a consequence of the second law of thermodynamics.

The inevitability of energy utilization with the inescapability of the dictates of the second law of thermodynamics throws up the challenge of not only energy utilization efficiency but also the sustainability of energy resource utilization. According to the famed Brundtland report, sustainable development is to "*ensure that humanity meets the needs of the present without compromising the ability of future generations to meet their own needs*" [1]. With the first industrial revolution in Europe (1800), coal became a major energy resource joined about a century later by petroleum. To maintain a good standard of living, modern society depends on many types of services, which conventionally, are rendered through combustion of fossil fuels. Therefore, although there are increasing drives toward alternative energy sources, there are also genuine needs to utilize the fossilized hydrocarbon fuels in as sustainable a manner as possible.

This chapter's contribution is thus about the metrics that can be used to assess the degree of sustainability of combustion of fossil carbon-based fuels. Exergy-based metrics are suitable for sustainability assessment of energy utilization due to the fact that, by definition, exergy combines both system and environmental properties. One definition of exergy is the maximum extractable work from a given system at a specified state till it attains thermodynamic equilibrium with the environment [2]. Thermodynamic equilibrium entails all manners of equilibria: mechanical, thermal, chemical, and so on, between the system and the environment. This is when all types of gradients: pressure gradient, temperature gradient, mass concentration gradient, and so on, cease between the system and the environment.

Conversely, for environmental harmony and ecological balance, a system should only release substances to the atmosphere in the 'dead state', when they are supposed to have zero exergy and zero disturbance in the environment. This is why exergy efficiency and its functions become very important in assessing energy and the environmental sustainability of energy utilization. This is because the more exergy efficient a process or system is, the less its exergy destruction level and the more its degree of sustainability [3].

Exergy reflects the quality of a resource, giving insight into which material or energy streams are worth recovering: streams with high exergy content have more potential for value extraction. Its foundation on the second law of thermodynamics provides an engineering understanding of the irreversibilities generated during production. One significant aspect of exergy efficiency is that it covers both energy and materials as a single indicator [4]. All other metrics considered in this chapter are functions of exergy efficiency, directly or indirectly.

Indeed, not only energy is degraded through entropy creation, but material resources are degraded as well [5, 6]. Besides, exergy destruction is synonymous with energy resource depletion. The easiest to understand is thermal energy emanating from combustion of hydrocarbon fuel. Some quantity of the fuel would be responsible for the degraded part of the energy or the destroyed exergy. The metric that directly takes care of material resource degradation is called depletion number.

Another metric, closely-related to depletion number, is the sustainability index. It is the reciprocal of depletion number and, therefore, a measure of resource conservation [7, 8]. The fourth metric in this chapter is exergetic improvement potential. When used in a system or process detailed analysis, it suggests feasible improvements in the system through the reduction of irreversibilities. The analysis starts with identification of areas of exergy destruction and possible technical ameliorations in the system within the limitations of the second law of thermodynamics [9].

Central to sustainable development is environmental protection. That is why there is a clamor for eco-friendly energy utilization. Hence, the fifth and last metric in the chapter is environmental compatibility. It measures the degree of eco-friendliness of a fuel. The major independent variable in the metric is the emission abatement exergy. Essentially, it compares the fuel's ordinary chemical exergy with its overall chemical exergy when the emission abatement exergy (as a result of its combustion) is factored in [10, 11].

2. Methodology

2.1 Exergy efficiency, ψ

This is defined by several authors [12, 13], as the ratio between the exergy delivered to the user, B_{out}, and the exergy input, B_{in}:

$$\psi = \frac{B_{out}}{B_{in}} \ldots (0 \leq \psi < 1) \tag{1}$$

Its lowest value is zero when the entire input exergy is destroyed. The highest value of unity is practically unattainable, due to the constraints imposed by the natural, second law of thermodynamics. Exergy efficiency is a dimensionless metric, sometimes expressed as a percentage.

For the cases considered in this chapter, the input exergy is always the fuel chemical exergy. The fuel chemical exergies have been found to be related to their heating values through an exergy factor [14]: φ_f. The fuel heating value, HV, its chemical exergy, ε_f, and the exergy factor are related as follows:

$$\varepsilon_f = \varphi_f HV \tag{2}$$

The fuel heating value can either be the higher heating value, HHV, or the lower heating value, LHV. When the system is multi-component, like in fuel-mix analyses, the overall system efficiency is given by:

$$\psi_{overall} = \frac{\sum_{i=1}^{n} \psi_i B_{in,i}}{\sum_{i=1}^{n} B_{in,i}} \tag{3}$$

$$= \sum_{i=1}^{n} \psi_i B'_{in,i} \tag{4}$$

In Eqs. (3) and (4), the "i" subscript refers to the i-th component. Eq. (4) is in terms of fractional exergy, B′ [15].

2.2 Depletion number, D_p

Depletion number is the ratio of the magnitude of exergy destruction within a system or process, B_d, to the magnitude of the exergy input [5]:

$$D_p = \frac{B_d}{B_{in}} \tag{5}$$

$$\text{But, } B_d = B_{in} – B_{out} \tag{6}$$

Hence, the depletion number is related to the system or process exergy efficiency thus:

$$D_p = 1–\psi \tag{7}$$

It is also a dimensionless metric, expressible as a percentage as well. As a metric, it is complementary to exergy efficiency.

Also, in a multi-component system,

$$D_{p,overall} = 1–\psi_{overall} = \frac{\sum_{i=1}^{n} D_{p,i} B_{in,i}}{\sum_{i=1}^{n} B_{in,i}} = \sum_{i=1}^{n} D_{p,i} B'_{in,i} \tag{8}$$

The lower the exergy efficiency, or higher the depletion number, the farther away from thermodynamic equilibrium the effluents from such a system or process and the more unsustainable it is.

2.3 Sustainability index, SI

SI is the reciprocal or multiplicative inverse of Depletion Number [16]. Hence,

$$SI = 1/D_p = \frac{1}{1-\psi} \tag{9}$$

The minimum value of SI is unity, when ψ is zero. It has no maximum value or upper bound. It is also a dimensionless metric. Since it is the reciprocal of depletion number, the practical indication of SI is clear: resource conservation. Hence, higher values (>1) indicate resource conservation and sustainability.

2.4 Improvement potential, IP

IP is an exergy-based sustainability metric [[17] in [13]] defined as:

$$IP = (1–\psi)(B_{in}–B_{out}) = (1–\psi)^2 B_{in} \tag{10}$$

In practice, IP is a function of two variables, namely, ψ and B_{in} (= $m\varphi_f HV$). It is a quadratic function of ψ (with a repeated solution of unity) and an increasing linear function of B_{in}, or fuel supply mass, m, since $\varphi_f HV$ is constant for any particular fuel.

It is non-dimensionless. Its unit is that of exergy. When ψ is zero, exergy is fully destroyed and IP is maximum at the value of the input exergy. The improvement challenge is then at its peak. It is instructive to note that system or process improvement through ψ increment is tantamount to reducing B_{in} (the fuel supply mass rate) at the same B_{out} level, thus conserving fuel material resources.

The importance of this metric is in the fact that it indicates the available opportunity to improve the system performance. This opportunity cannot be higher than the fuel chemical exergy at 100% exergy efficiency. Hence, the fuel high chemical exergy is an opportunity that can be harnessed with high system or process exergy efficiency.

2.5 Environmental compatibility, ξ

Environmental compatibility of energy utilization is a measure of energy resource sustainability, defined [10, 11] as the ratio of the net input fuel exergy to its gross exergy when combustion emission abatement exergy is added. When the emission abatement exergy is independently supplied (from a non-Carbon-based fuel source), the environmental compatibility is:

$$\xi = \frac{B_{in}}{B_{in} + B_{EA}} = \frac{1}{1 + \frac{B_{EA}}{B_{in}}} = \left(1 + \frac{B_{EA}}{B_{in}}\right)^{-1} \tag{11}$$

B_{EA} is the emission abatement exergy.

If the abatement exergy is low, then ξ is high. The abatement exergy is to offset the ecological imbalance produced as a result of exergy destruction, manifesting as emissions. Mathematically, the value of ξ depends on the ratio of B_{EA} value to that of B_{in}. It is also dimensionless.

In the alternative case, when emission abatement exergy is obtained from the supplied fuel exergy, the following expression is used for environmental compatibility, ξ [18]:

$$\xi = \frac{B_{in} - B_{EA}}{B_{in}} = 1 - \frac{B_{EA}}{B_{in}} \tag{12}$$

Both equations produce the same maximum value of unity (100%), when the abatement exergy is nil. Besides, a binomial expansion of Eq. (11) converges to Eq. (12) for small values of the ratio B_{EA}/B_{in} (x):

$$\xi = \left(1 + \frac{B_{EA}}{B_{in}}\right)^{-1} = 1–x + x^2–x^3 + x^4–x^5 + ... + (–x)^n \tag{13}$$

$|x| \leq 1$; $x = B_{EA}/B_{in}$.

Dewulf et al. [19] quoted Dewulf et al. [10] that the abatement exergy of CO_2 is 5.86 MJ/kg. Cornelissen [9], as quoted by Dewulf et al. [19], obtained abatement exergy values for SO_x, NO_x and phosphate as 57, 16 and 18 MJ/kg, respectively. Also, in Tang et al. [18], the abatement exergy values for CO_2, SO_2 and NO_x were found to be 5.9, 57.0 and 16.0 MJ/kg, respectively. Hendriks [20], referenced by De Swaan Arons [21], reported that 5.862 MJ abatement exergy would be required per kg CO_2 produced from non-renewable energy sources as well.

For instance, to obtain CO_2 abatement exergy of a fuel (MJ/kg of fuel), we multiply chemical exergy value of the fuel consumed (MJ/kg of fuel) by the effective CO_2 emission factor (kg of CO_2/MJ of fuel, in **Table 1**) to obtain the mass of CO_2 produced (kg per kg of fuel), and then multiply by 5.862 MJ per kg of CO_2. This procedure was followed to obtain the data in **Table 2** from the ones in **Table 1** and Garg et al. [22]. Data in **Table 2** formed the basis for **Figure 1**.

In **Table 2**, effective CO_2 emission factor, (A $\times$ 44/12), is mass of CO_2 produced per MJ of fuel used.

The emission factor value for blast furnace gas includes carbon dioxide originally contained in this gas as well as that formed due to combustion of this gas. Blast furnace gas produces the highest quantity of CO_2 per unit of fuel exergy utilized [22].

Fuel	LHV (MJ/kg)	Exergy factor
LPG	47.3	1.056
Kerosene	43.8	1.07
Diesel Fuel	43.0	1.07
Gasoline	44.3	1.07
Natural Gas	48.0	1.04
Residual Fuel Oil	40.4	1.07
Blast Furnace Gas	2.7	1.06

Table 1.
Selected fuel data [22, 23].

Fuel	Chemical exergy, ε_f (B_{in}) (MJ/kg)	Default carbon content, A (kg/MJ of fuel)	Effective CO_2 emission factor, A × 44/12	CO_2 mass per mass of fuel (kg/kg)	B_{EA} (MJ/kg of fuel)	B_{EA}/B_{in}
LPG	49.9488	0.0172	0.0631	3.1518	18.4757	0.3699
Kerosene	46.866	0.0196	0.0719	3.3697	19.7530	0.4215
Diesel Fuel	46.01	0.0202	0.0741	3.4093	19.9856	0.4344
Gasoline	47.401	0.0189	0.0693	3.2849	19.2560	0.4062
Natural Gas	49.92	0.0153	0.0561	2.8005	16.4166	0.3289
Residual Fuel Oil	43.228	0.0211	0.0774	3.3458	19.6134	0.4537
Blast Furnace Gas	2.808	0.0708	0.2596	0.7290	4.2731	1.5218

Table 2.
Chemical exergy and CO_2 data on selected fuels [22, this work].

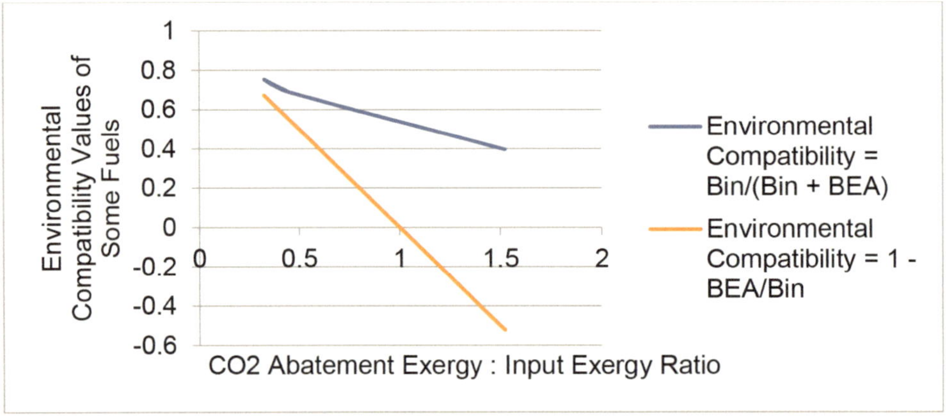

Figure 1.
The two environmental compatibility functions of B_{EA}/B_{in} for CO_2 emission for selected fuels in practice.

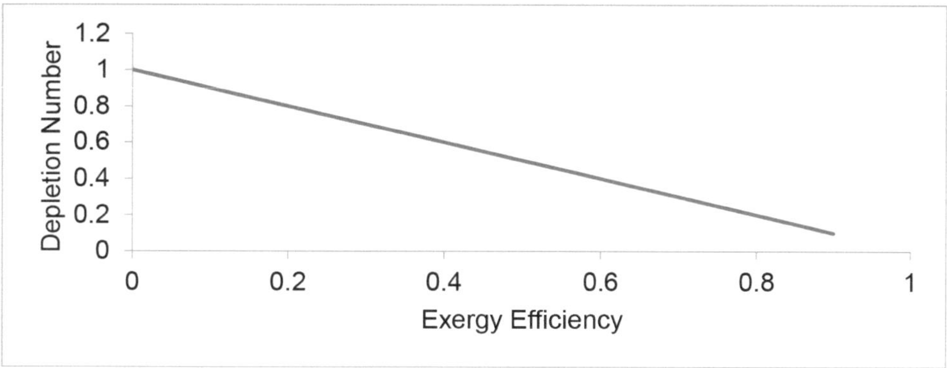

Figure 2.
Depletion number as a function of exergy efficiency.

3. Results and discussion

3.1 Exergy efficiency

Lowest value of exergy efficiency is zero, as total exergy destruction is possible. However, a maximum value of 100% for exergy is unattainable, due to the constraints imposed by the second law of thermodynamics. These two facts affect the extreme values of other metrics which are direct functions of system or process exergy efficiency. Apart from environmental compatibility, exergy efficiency is an independent variable of other metrics considered in this chapter. Indeed, it is the sole independent variable of depletion number and sustainability index and one of the two independent variables of improvement potential.

3.2 Depletion number

D_p ranges from unity, when ψ is zero, to close zero when ψ is very high. Hence, its maximum value is unity. However, it can never be zero, since ψ can never be unity. It is a decreasing, linear function of ψ, approaching zero as ψ approaches unity, as shown **Figure 2**. The smaller its value is, the better. For an efficiency of 40%, for instance, the depletion number is 0.60.

3.3 Sustainability index

It is an increasing function of ψ, with no defined upper bound, as shown in **Figure 3**. It can also be seen in **Figure 3** that SI seems to increase astronomically beyond an exergy efficiency value of 0.9. This observation is brought out more vividly in **Figure 4**.

Indeed, a binomial expansion of Eq. (9) gives SI as a monotonously increasing power function of exergy efficiency, sum of a geometric progression with common ratio of ψ:

$$SI = 1 + \psi + \psi^2 + \psi^3 + \dots + \psi^{n-1} \; (0 \leq \psi < 1) \tag{14}$$

The way SI increases in **Figures 3** and **4** is not surprising, considering Eq. (14). **Figure 4** is a plot of the first derivative of **Figure 3**, and (considering Eq. (14)), is still

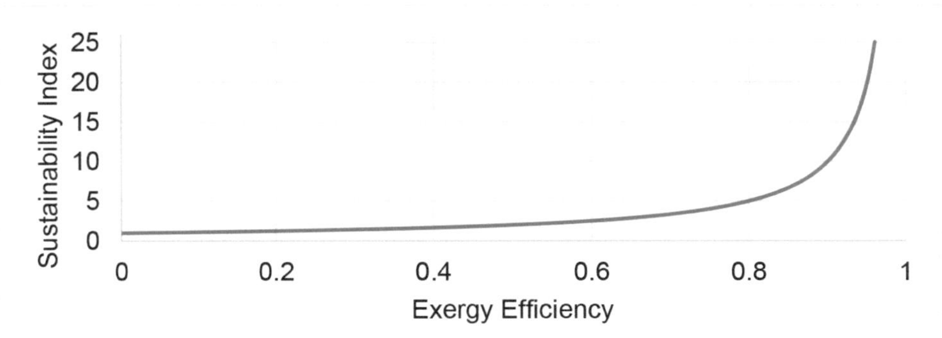

Figure 3.
Sustainability index as a function of exergy efficiency.

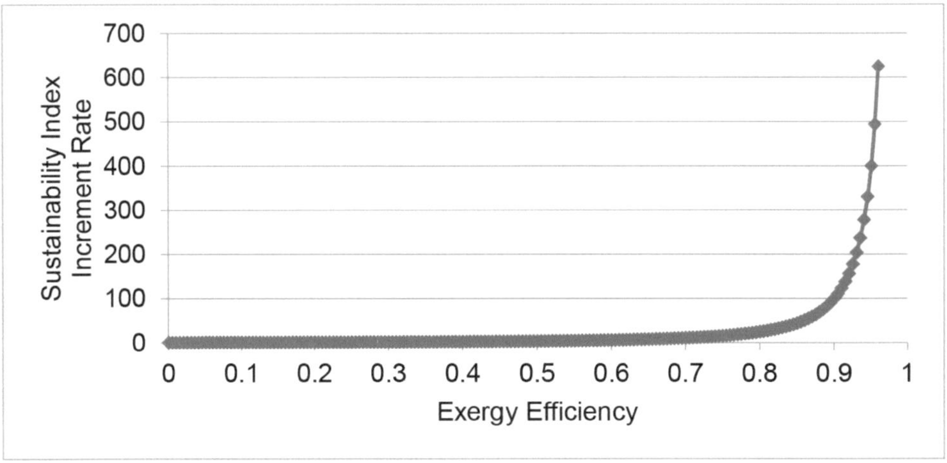

Figure 4.
Sustainability index increment rate.

a plot of a power function of ψ. However, as the power of ψ comes down, the value goes up, since $\psi < 1$. This is why the scale on the vertical axis of **Figure 4** is higher than that of **Figure 3**. When exergy is fully destroyed and ψ is zero, SI attains its minimum value of unity. At any other attainable value of ψ, it (SI) keeps on increasing, being a power function of ψ. Hence, any favorable value of SI must be very high. For instance, for an efficiency of 40%, SI is 1.67; whereas if the efficiency is doubled (80%), SI becomes 5.0.

3.4 Improvement potential

Improvement Potential is a function of two variables: exergy efficiency and input exergy. However, for a particular fuel (fixed chemical exergy per unit mass), IP has only ψ as a variable, as shown **Figure 5**. It is also shown in **Figure 5** that the potential decreases as the efficiency is improved. This is expected to be done through process or system performance improvement. Since the potential is exhausted at the highest attainable exergy efficiency, a further opportunity is only possible at a higher value of

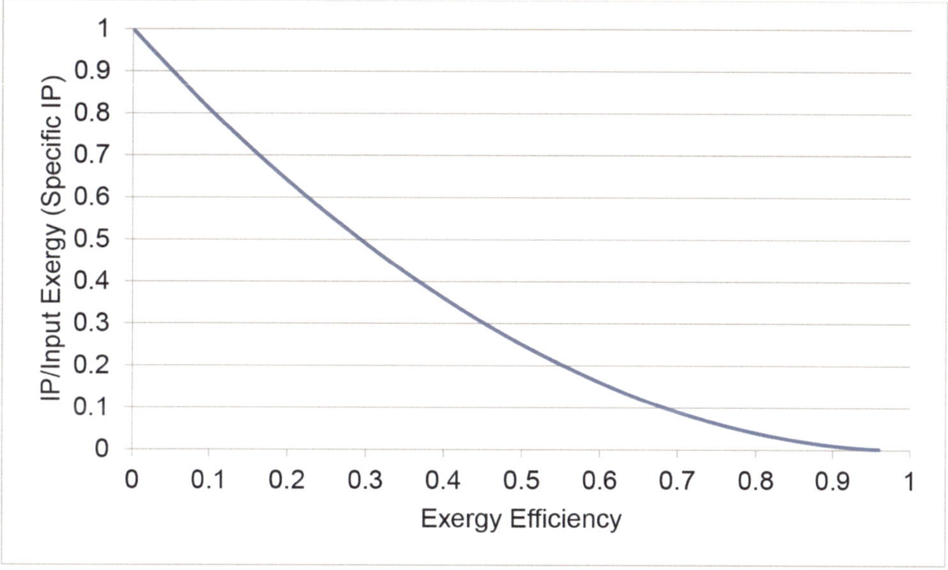

Figure 5.
Specific improvement potential as a function of exergy efficiency.

chemical exergy. Noting that chemical exergy is a fuel property, the practical implication of this further potential enhancement is fuel substitution. **Figure 6** explains this fact further graphically by comparing different fuels based on their chemical exergies.

There, it is seen that, at a particular value of system or process efficiency, the residual fuel oil has a lower improvement potential than natural gas. In other words, for the same quantity of fuel consumption, a higher efficiency can be attained with improved process or system technology with natural gas, than with residual fuel oil.

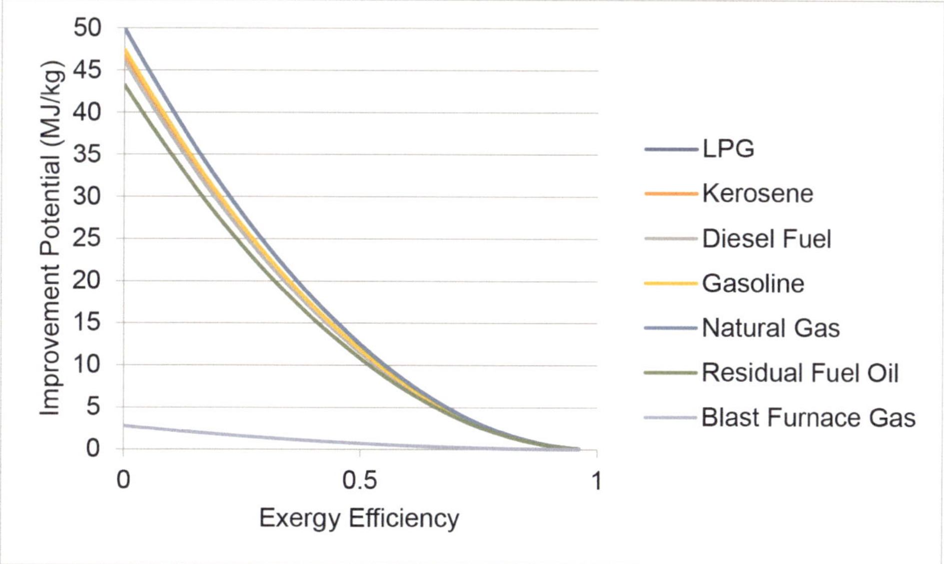

Figure 6.
Improvement potentials of fuels as functions of exergy efficiency.

It is well known that gas fuel utilization is, in general, more sustainable that liquid fuel utilization. An exception to this is a fuel like the Blast Furnace gas, which has a very low chemical exergy.

3.5 Environmental compatibility

Like the previously mentioned metrics, environmental compatibility is a function of a ratio of two exergies. But unlike them, it is not a function of exergy efficiency. Rather, it is a function of the ratio of abatement exergy to the input exergy, as can be seen in **Figure 7** for CO_2 emission. High carbon fuels have high abatement exergies and low environmental compatibilities. It is easily understood that if other emissions like SO_2 and NO_x are considered, fuels with high abatement exergies like high sulfur fuels would have low environmental compatibilities. The fact that the independent variable on the abscissa of **Figure 7** is invariably a property of a fuel suggests that fuel substitution is a strong factor to be considered in improving environmental compatibility of a process or system. However, it is also noteworthy that system engineering maintenance is a factor too. A properly and promptly maintained system will definitely be more environmentally compliant, even with same fuel type.

The fact that Eqs. (11) and (12) are about two different approaches to measure the parameter is responsible for the different values obtained, especially at higher exergy ratios (**Figures 1** and 7). Besides, the fact that environmental compatibility values obtained through Eq. (11) are higher than those obtained through Eq. (12) is also logical. The emission abatement exergy in Eq. (11) is obtained through non-carbon-based sources, while the one in Eq. (12) is obtained through the same carbon-based sources. It is also expected that Eq. (12) would yield a negative value for environmental compatibility for a fuel like blast furnace gas with B_{EA} more than B_{in} (**Figure 1**).

It is indeed outrageous and unsustainable for B_{EA} to be close to, equal to, or more than B_{in}. This is because the practical implication is that the entire input fuel chemical exergy is used to abate its combustion emissions. In this case, there will no net gain from the fuel combustion process. Hence, both equations are equivalent and applicable under practical, sustainable situations. Under this situation, the minimum value of environmental compatibility is zero, using Eq. (11) or Eq. (13).

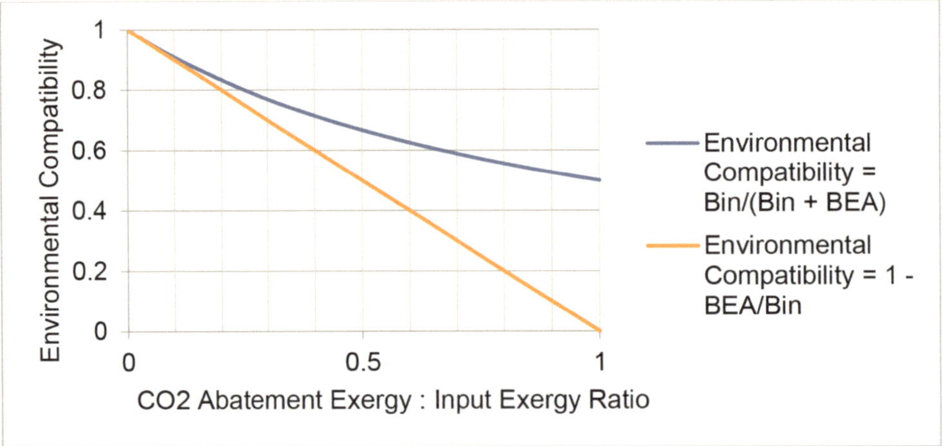

Figure 7.
The two general environmental compatibility functions of B_{EA}/B_{in} for CO_2 emission in principle.

4. Conclusion

Five exergy-based metrics of sustainable energy utilization have been considered in this chapter. They are exergy efficiency, depletion number, sustainability index, improvement potential and environmental compatibility. The first three metrics are functions of exergy efficiency. They can always be influenced for improvement through ingenuous process path or system technology design. The fourth one is a function of both exergy efficiency and input exergy. Being a function of both exergy input and efficiency, its improvement can be effected through fuel substitution as well as process path and system technology design. The fifth metric is a function of the ratio of emission abatement exergy to the input exergy. Since the emission abatement exergy is a strong function of the fuel, its improvement is mainly dependent on fuel substitution. However, all of them are separate functions of system condition, and may therefore be improved through prompt and adequate system maintenance. They all underscore the importance of exergy in sustainability analyses and eco-friendliness of energy utilization.

Author details

Ismaila Badmus
Mechanical Engineering Department, Yaba College of Technology, Lagos, Nigeria

*Address all correspondence to: ismaila.badmus@yabatech.edu.ng

References

[1] World Commission on Environment and Development. Our Common Future. Oxford, UK: Oxford University Press; 1987. p. 383

[2] Kanoglu M, Dincer I, Cengel YA. Exergy for better environment and sustainability. Environment, Development and Sustainability. 2009; **11**:971-988. DOI: 10.1007/s10668-008-9162-3

[3] Rosen MA. Exergy analysis as a tool for addressing climate change. European journal of sustainable development. Research. 2021;**5**(2):em0148. e-ISSN: 2542-4742. DOI: 10.21601/ejosdr/9346

[4] Hernandez AG, Cullen JM. Exergy: A universal metric for measuring resource efficiency to address industrial decarbonisation. Sustainable Production and Consumption. 2019. DOI: 10.17863/CAM.42730

[5] Connelly L, Koshland CP. Two aspects of consumption: Using an exergy-based measure of degradation to advance the theory and implementation of industrial ecology. Resources, Conservation and Recycling. 1997;**19**:199-217

[6] Georgescu-Roegen N. The entropy law and the economic process in retrospect. Eastern Economic Journal. 1986;**12**:3-25

[7] Li B, Li C, Li T, Zeng Z, Ou W, Li C. Exergetic, energetic, and quality performance evaluation of Paddy drying in a novel industrial multi-field synergistic dryer. Energies. 2019;**12**(23): 4588. DOI: 10.3390/en12234588

[8] Bayat M, Özalp M. Sustainability assessment of a solar photovoltaic module with exergetic efficiency approach. In: Proceedings of the 9th International Exergy, Energy and Environment Symposium (IEEES-9), 14-17 May 2017; Split, Croatia. 2017. pp. 1-8. Available from: http://www.ieees9.fesb.unist.hr/

[9] Cornelissen RL. Thermodynamics and Sustainable Development. [PhD thesis]. Enschede: Universiteit Twente; 1997

[10] Dewulf J, Van Langenhove H, Mulder J, van den Berg MMD, van der Kooi HJ, de Swaan AJ. Illustrations towards quantifying the sustainability of technology. Green Chemistry. 2000;**2**: 108-114

[11] Magnanelli E, Berglihn OT, Kjelstrup S. Exergy-based performance indicators for industrial practice. International Journal of Energy Research. 2019;**Preprint**:1-25

[12] Rosen MA, Dincer I, Kanoglu M. Role of exergy in increasing efficiency and sustainability and reducing environmental impact. Energy Policy. 2008;**36**:128-137

[13] Hammond GP. Industrial energy analysis, thermodynamics and sustainability. Applied Energy. 2007;**84**: 675-700. DOI: 10.1016/j.apenergy.2007.01.002

[14] Szargut J, Morris DR, Steward FR. Exergy Analysis of Thermal, Chemical, and Metallurgical Processes. Washington DC, USA: Hemisphere Publishing Corporation; 1988

[15] Badmus I. Sustainable energy analysis of Nigerian road transportation sector: Benchmarking against global trends. Journal of Energy Research and Reviews. 2021;**9**(4):1-20. Article no: JENRR.79560. ISSN: 2581-8368. DOI: 10.9734/JENRR/2021/v9i430238

[16] Dincer I, Zamfirescu C. Sustainability dimensions of energy. Comprehensive Energy Systems. Elsevier Inc. 2018;**1**: 101-152. DOI: 10.1016/B978-0-12-809597-3.00104-8

[17] Van Gool W. Exergy analysis of industrial processes. Energy. 1992;**17**(8): 791-803

[18] Tang Y, Dong J, Li G, Zheng Y, Chi Y, Nzihou A, et al. Environmental and Exergetic life cycle assessment of incineration- and gasification-based waste to energy systems in China. Energy, Elsevier: 1 – 10. 2020;**205**: 118002. DOI: 10.1016/j.energy.2020.118002ff. hal-02863924ff

[19] Dewulf J, Van Langenhove H, Dirckx J. Exergy analysis in the assessment of the sustainability of waste gas treatment systems. The Science of the Total Environment. 2001;**273**:41-52

[20] Hendriks C. Carbon dioxide removal from coal-fired power plants [PhD thesis]. Netherlands: University of Utrecht; 1994

[21] De Swaan AJ, van der Kooi H, Sankaranarayanan K. Efficiency and Sustainability in the Energy and Chemical Industries. New York, NY: Marcel Dekker; 2004. p. 300

[22] Garg A, Kazunari K, Pulles T. Introduction. In: 2006 IPCC Guidelines for National Greenhouse Gas Inventories, Volume 2: Energy. Chapter 1. 2006. pp. 1-29

[23] Badmus I, Bakri AJ. Energy, exergy and environmental compatibility analyses of LPG and household kerosene utilisations as domestic fuels in Nigeria: 1980–2019. Journal of Energy Research and Reviews. 2021;7(3):46-56. Article no: JENRR.67464. ISSN: 2581-8368. DOI: 10.9734/JENRR/2021/v7i330194